Praise for *Pawpaws*

Blake Cothron's book makes a much-needed contribution to the pursuit of pawpaw culture. It seems that interest in pawpaws has exploded beyond my imagination since the year 1976 when I undertook my quest to bring pawpaws out of the shadows of the forest and into the sunlight of gardens and orchards. This enthusiasm is a good thing, and it now requires answers to the many questions that naturally arise for those new to pawpaws. And since pawpaws are a new crop, there are many questions that orchardists and hobbyists have that Blake expertly addresses. The future will hold more questions and definitely the need for more breeding—an opportunity for the patient and educated amateur.

— R. Neal Peterson, founder, Peterson Pawpaws,
agricultural economist, USDA Economic Research Service

Planting a single pawpaw is a re-evolutionary act. Mixed use forests once blanketed us. Acorns, chestnuts, pawpaws, and their regional equivalents were the princes of that realm. With this valuable book, you can pawpaw your own food forests, restoring the diversity, abundance, and climate we all need.

— Albert Bates, permaculture instructor, ecovillage designer,
author, *The Biochar Solution: Carbon Farming and Climate Change*

Blake Cothron is an authority on pawpaws and provides a clear, detailed guide for commercial success in growing this "oddly appealing species" (his own words). The supply of this exotic, trending, easy-to-grow fruit has not yet met the demand. Blake shares the wealth of his knowledge, including challenges—and when he doesn't know, he says so (it's probable that others don't know either).

— Pam Dawling, author, *The Year-Round Hoophouse*
and *Sustainable Market Farming*

My own fascination with pawpaws began nearly 40 years ago with a visit to pawpaw breeder Tom Mansell's Paw Paw Haven near Pittsburgh PA. Mr. Mansell's food forest landscape had twenty-two varieties of pawpaws at one point. Eventually I discovered my own secret patch. Each September for the past twenty-five years I have waded through tall weeds, navigated a swampy stream bed, and traversed a small woodlot to harvest a few dozen complexly flavored fruits from my secret patch. These I share with friends and family and introduce new people to this uncommon seasonal delight. The pawpaw's revival is long overdue. Blake Cothron's *Pawpaws* will help bring about the day when fragrant fruit is no longer a rare treat, but a regular part of our seasonal diet.

— Darrell E. Frey, Three Sisters Farm, author, *Bioshelter Market Garden*, co-author, *The Food Forest Handbook*

Pawpaws

THE COMPLETE **GROWING** AND **MARKETING** GUIDE

BLAKE COTHRON

Cover design by Diane McIntosh.
Lower middle, lower right © Blake Cothron. Others © iStock

All photographs copyright © Blake Cothron unless otherwise noted.

Printed in Canada. First printing May 2021.

This book is for educational and informational purposes only. The trade names of certain products are for convenience and reference only. No special endorsement of a particular product or infringement with any trade name is intended. Products not mentioned but with equal effectiveness could be used in successful production. It is not our intent to criticize any products not mentioned (though we do only advocate organic products). Remember that these are provided as guidelines only and are not a substitute for product labeling, instruction, and manufacturer information. All agriculture is done at your own risk, and the author and publisher are not responsible for any loss or damages due to following the advice and recommendations in this book.

Inquiries regarding requests to reprint all or part of *Pawpaws* should be addressed to New Society Publishers at the address below. To order directly from the publishers, please call toll-free (North America) 1-800-567-6772, or order online at www.newsociety.com

Any other inquiries can be directed by mail to:

New Society Publishers
P.O. Box 189, Gabriola Island, BC V0R 1X0, Canada
(250) 247-9737

LIBRARY AND ARCHIVES CANADA CATALOGUING IN PUBLICATION

Title: Pawpaws : the complete growing and marketing guide / Blake Cothron.

Names: Cothron, Blake, 1985– author.

Description: Includes bibliographical references and index.

Identifiers: Canadiana (print) 20200396447 | Canadiana (ebook) 20200396889 | ISBN 9780865719552 (softcover) | ISBN 9781550927481 (PDF) | ISBN 9781771423441 (EPUB)

Subjects: LCSH: Pawpaw. | LCSH: Pawpaw—Marketing.

Classification: LCC QK495.A6 C68 2021 | DDC 634/.41—dc23

Funded by the Government of Canada | Financé par le gouvernement du Canada | Canada

New Society Publishers' mission is to publish books that contribute in fundamental ways to building an ecologically sustainable and just society, and to do so with the least possible impact on the environment, in a manner that models this vision.

Contents

Preface ix
Introduction 1
1. Foraging for Wild Pawpaws 13
2. Description of North American Pawpaw Fruit 17
3. Flowering and Pollination 29
4. Site Design and Planting 43
5. Choosing Your Trees 59
6. Maintaining the Orchard 69
7. Harvesting Pawpaw Fruit 93
8. Tree Propagation 97
9. Pests, Diseases, Disorders, and Their Management 107
10. Pawpaw Fruit Marketing Strategies 131
11. Pawpaw Cultivars 151
12. Troubleshooting, Cost Analysis, and Calendar 183
13. Conclusion 195
Resources 197
Notes 199
Index 203
About the Author 211
About New Society Publishers 212

This book is dedicated to Bhagavan Sri Krishna,
Who likes to sit under beautiful flowering and fruiting trees
and play His flute. May I always hear and follow that flute song.

I pray that this book will help empower people to plant more trees,
feed people, grow organic food, and do good things in their communities.

Preface

It's been said that agriculture is the most noble profession. Horticulture is, to me, an incredible art and science meant to optimize and refine agriculture by developing effective growing techniques and even by refining plants themselves. I've been intrinsically drawn to it and deeply involved in it most of my life and bring with it a deep love and sensitivity to nature and all natural systems. I'm humbled to help move horticulture forward in any little small way I can by writing this book and operating our five-acre Certified Organic research and production farm, Peaceful Heritage Nursery, in Stanford, Kentucky.

The nearly four years spent writing this book have been mostly easy and enjoyable because I felt the topic was fairly simple and straightforward. After writing a couple of hundred pages, I realized the topic is actually much more complicated than first estimated, but I am pleased with how the book has turned out, and I hope readers will find it useful and easy to read. Also, in my course of writing, I realized there are a lot of missing links and unknowns in the world of pawpaws that I focused on connecting and helping uncover.

I became very attracted to the North American pawpaw tree (*Asimina triloba*) as a teenager in the early 2000s and have had a strong affinity for this oddly appealing species since then. Beginning around 2009, I found cultivating the trees mostly easy and quite fun, especially since I was growing them in an absolutely ideal situation on perfect pawpaw soil and climate in Louisville, KY.

In our 21st century Western civilization, we love to complicate basic matters and to make technical what should actually be pretty simple. In many non-Western cultures, if you want a fruit tree, you simply throw the seeds out or plant them in your garden and wait. If it gets fruit, then good. If it gets a lot of fruit, then sell some. If it dies, plant another one. Our Western preoccupation with (attempting to) control nature and our obsession with the future, maximizing profitability, market projections, analyzing possible scenarios, etc. bleeds into our horticultural ventures and gardening, and I'm here to say: just relax a little. Pawpaws are pretty easy to grow, especially compared to highly bred and human-pampered/tampered fruits like peaches and apples that wilt at the sight of insects and diseases. Pawpaw is still a furry chested, burly, backwoodsman of a tree and has not succumbed to the manicured, air-conditioned, civilized hybridization process thus far. Although, truth be told, as we will learn later, it is quite fragile compared to other native species such as oaks, poplars, and maples.

There are some people out there who helped and contributed in some way to make this book happen. These are, in no order of importance: Ron Powell, president of the Ohio Chapter of the North American Pawpaw Growers Association (NAPGA) who contributed hours of phone dialogue about pawpaws, as well as being very generous with sharing his notes and information, and helping to edit the manuscript; R. Neal Peterson, who pioneered the commercialization of pawpaws and developed some of the best modern cultivars, and whose work has contributed much to our ongoing knowledge of pawpaw culture. His work is quoted or referenced a number of times throughout this book, but truly his contribution has been so large and unanimous that I am sure I unknowingly quote information here and there that is derived from his work somehow or other. Neal also was very generous with his time and helped to edit the manuscript, which I assure you he did quite thoroughly. Cliff England, Kentucky nurseryman, tireless fruit explorer as well as passionate pawpaw and unusual fruit connoisseur, taught me just how good jujube fruits actually are, as well as how to graft pawpaws, and has always been very generous with his knowledge and time. Next there is Ms. Sheri Crabtree, who took the time to very carefully help edit the final manuscript, as well as Jeremy Lowe, Kirk Pomper and the other pawpaw program research staff at Kentucky State University (KSU), whose pawpaw work is inspiring growers around the world, and whose work and photographs contributed much to this book, as well as our collective pawpaw knowledge base. And, last but certainly not least, the late, great Jerry Lehman, whose pawpaw breeding and cultivation work was impressive and inspiring (as well as consistently award-winning). I would be deficient in many areas, including agriculturally, if not for my wife, Rachel, who has been with me for years, not only tolerating my fleeting obsessions but also annually extracting hundreds of pounds of pawpaw seeds from slimy pulp and grafting hundreds of pawpaw trees by my side.

My aim with this book was to create a highly readable, enjoyable, and very intensive exploration into the cultivation of North American pawpaw. I hope it will prove a very practical and useful guide to help novices and experienced growers alike to have good success growing this amazing tree for fun, food, profit, or all three.

So, with that let's get into it and cover the basics.

Introduction

In the shady hollows and backwoods stream banks, in parks and wooded edges of farmland across much of eastern America, a graceful, humble little tree quietly grows that happens to produce the largest edible fruit in North America. This fruit is traditionally known as pawpaw, or, in botanical language, *Asimina triloba*. It's a hot topic right now among many diverse circles: horticulture, botany, gourmet chefs, market farmers, brewers, even medical researchers looking into the potential uses of the tree in cancer treatment. Interest in cultivating North American pawpaw is increasing rapidly in many temperate countries around the world, including Japan, South Korea, France, Germany, Ukraine, Austria, Australia, Germany, and Italy. Why is that? The answer is multifaceted: pawpaw trees are extremely cold-hardy and adaptable to a range of temperate growing conditions, perform well without the addition of chemical sprays, and produce a fruit that is exotic enough that you'd think it came out of the Amazon. The fruit also happens to be very delicious, nutritious, large size (up to one pound or more each), and produced in abundance when the trees are well grown. Farmers, processed food manufacturers, nurserymen, and chefs are always looking for an exciting, lucrative, and trendy new product, and pawpaw fits the bill quite well. So far, its elusive presence and short season have created a lot of mystique (but that may be set to change soon).

Even though currently you will not find pawpaws in nearly any standard US grocery store, they are becoming more commonly seen in many late-summer farmer's markets and health food stores; there is even demand for mail-order fresh pawpaw fruit. Still, the question remains, why are pawpaws not found in stores and why are they so extremely rare? The answer basically boils down to lack of information, growers, and education; there's not a lot of resources and in-depth information for farmers and marketers to study to see if pawpaws might be a good product for them to grow and sell. The pawpaw industry in general is also brand-new, and so far, nursery production of high-quality trees has also been extremely limited.

Although demand for the fruit and the trees is rapidly increasing, as of this writing there has not been a *single book* written that describes the use, marketing, and cultivation of pawpaws in *full detail* aimed to assist commercial and market growers. Mostly the only resources have thus far been a small number of university publications, brief online articles, glossy hypes in magazines, and (usually very brief) mentions of pawpaws in various fruit-growing books. The focus of most of the current information has been on describing the fruit itself and explaining how to grow a few backyard pawpaw trees. Any information geared toward farmers has been relatively short and not comprehensive. In the past, and sometimes still, pawpaws were considered so minor as barely worth mentioning as a mere novelty species with a few named varieties out there in the nursery trade somewhere. A short book about cultivating pawpaws was published years ago in Canada, but the book was not widely printed, not applicable in much of the US, and is very difficult to acquire. This book sets out to change that situation and to provide a comprehensive, detailed, yet easy and fun-to-read book for the market farmer, rare-fruit grower, backyard grower, and potential pawpaw farmer alike.

I met my first pawpaw around 2003 at church camp in rural central Indiana. One afternoon while walking along the edge of the woods near the pond, I came across a curious small tree upon which hung an odd, lumpy green fruit cluster. Intrigued, I asked one of the camp counselors if she knew what this was. The counselor, being a rural Hoosier herself, knew right away, and told me (the suburbanite teenager) that these were called "pawpaws," and they were in fact edible and even tasty. My question was answered, but my curiosity was thus piqued. That following autumn, I got a few close friends together, and bent on getting some pawpaw fruit, we were soon rummaging through the bushes and trees of our local Cherokee Park in Louisville, Kentucky, trying our darnedest to locate even one ripe fruit. Pawpaws actually grow abundantly in this park, so we were finding fruit, but it was still rock hard and underripe. At the tail end of our several hours' long journey through the backwoods of the park, a little discouraged and about ready to give up, I found a tree with a few golf ball-sized fruits dangling from the branches. You can imagine my excitement when, upon squeezing them, I discovered they were soft and ripe for the picking! We all feasted on them right there on the spot. They were sweet, soft, and delicious, with big, hard, shiny black seeds. After that experience, I was sold: pawpaws were really cool!

Now, many years later, pawpaws are a hot topic, in part due to their "exotic yet local" mystique, luscious tropical flavor profile, raving

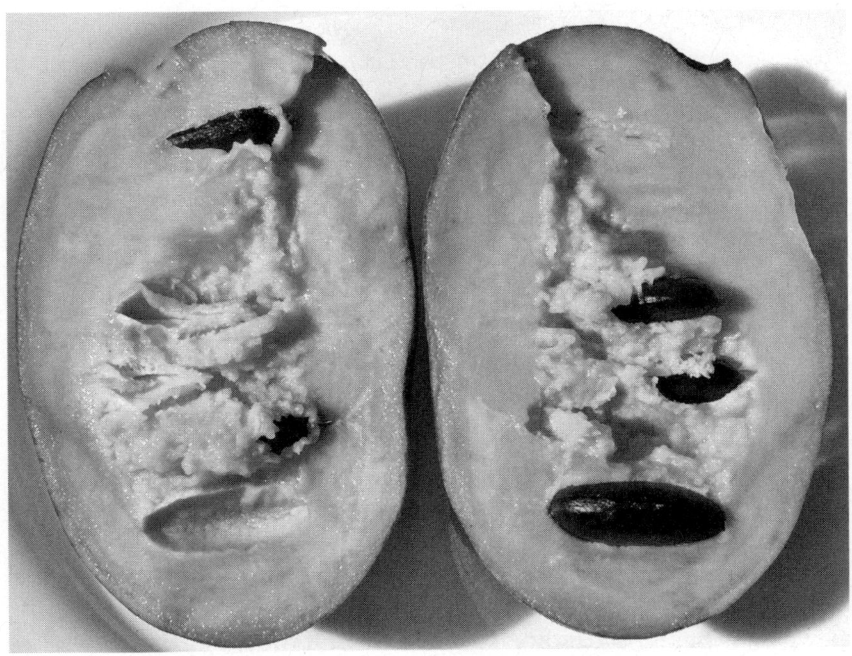

'KSU Benson' pawpaw fruit cut in half.

reviews, and fleeting annual appearance and disappearance. People want pawpaws. Growers want to grow pawpaws. Marketers want to market pawpaws. Chefs want to cook with pawpaws. Brewers want to brew with pawpaws. However, there is scant fruit supply and historically very little information to help make this happen. This book contains the in-depth information that eco-conscious and organic growers need. I believe organic agriculture is the future of most agriculture, and pawpaws being so easy to grow organically, why not do it this way anyway?

Indeed, there's something oddly curious and mysterious about the North American pawpaw, including a strange effect they seem to have on certain individuals. Some develop such an intense liking for pawpaws that they become obsessively engrossed in the growing, propagating, and eating of the fruit. Others are off-put by them and want nothing to do with them in any form. Pawpaws seem to create both diehard fans and dissenters. We'll explore why this is, and how to facilitate more positive pawpaw experiences, especially for marketing purposes. If you're reading this book, you want to know how to grow great pawpaws and possibly how to market them. Therefore, the main purpose of this book is to encourage and empower people who wish to grow pawpaw trees and educate them to do so effectively, including marketing fruit profitably if that is their goal.

This book emphasizes choosing excellent cultivars, proper siting of an orchard, effective planting, seasonal tree maintenance, and harvesting. Pollination, fertilization, insect, pest and disease control, as well as some basic nursery propagation practices are all covered in detail.

What Is a Pawpaw?

If you are reading this book, you probably already know exactly what a pawpaw is. But, in case you do not or seek to learn more, the pawpaw is a tree species categorized in botany by the Latin name *Asimina triloba*, pronounced "Ah-sim-ih-nuh try-lobe-uh." We'll start with the pawpaw's current status and briefly go back in time to review its colorful past.

Pawpaws, until recently an overlooked small tree quietly growing in the wilds of the understory forest, are currently being bred and developed to be more appropriate and marketable for commercial fruit growing. This breeding and improvement is being conducted privately, as well as by at least one university (Kentucky State University), with the goals of producing bigger, better-tasting fruits with fewer seeds on highly productive trees that resist disease. There are also other goals, such as breeding fruits low in acetogenins, a compound found in *Asimina triloba* that is very bioactive and not currently well understood (more on that later). Wild pawpaw fruits tend to be small, weighing only a few ounces, nearly half being large seeds, making the fruit not very marketable. Currently, pawpaw fruit is showing up more often in farmer's markets, trendy local food stores, gourmet restaurant menus, micro-brewed beers, wines, jams, and specialty ice cream. They are generally considered a hot item when available in their brief peak season, mostly within the areas where they are grown and recognized. By the way, pawpaw ice cream is *heavenly*.

Healthy pawpaw foliage in early summer.

Pawpaws have a rich but quiet history among rural Americans, being enjoyed by many a backwoods child as well as the millions of deer hunters (who of course spend a lot of time in the woods in September and October, and know a good deer bait tree). In the early 1900s, they were sold in many local produce markets, but gradually slipped into near obscurity, pushed out of the picture by more marketable and useful fruits with better storage capabilities, such as apples. Eventually pawpaws became something that practically only rural dwellers appreciated, and usually quietly. Before that, pawpaws were enjoyed for millennia by eastern Native Americans, and later early American colonists, explorers, and settlers.

The journal of Lewis and Clark tells us that the expedition was saved from starvation and death by Sacajawea and possibly other Native Americans who taught the explorers how to utilize the high-protein and high-sugar pawpaw fruits dropping along the wilderness trail.[1] The Native Americans knew the pawpaw quite well because they had eaten and appreciated them for millennia. The Native Americans were the first to cultivate the fruit in plots, likely finding good trees worthy of growing and planting the seeds, but also perhaps transplanting the young trees. Once I heard that archaeologists sometimes locate ancient Native American village sites by finding an abundance of pawpaw trees all within one isolated area, because they not only grew the fruit but also used the large shiny black seeds for game pieces, and some of these no doubt sprouted and took root. Thus the village became inundated with pawpaw trees, forming a nice symbiotic relationship with their human friends and benefactors.

The tropical-looking tree (and fruit) itself almost seems out of place in the temperate zones where it is native. Some people theorize that the pawpaw migrated gradually north from Central America, either on its own via the guts of animals such as deer, and even mastodons and giant ground sloths (how cool does that sound?), or by the work of Native peoples, who gradually planted it further and further north, and thus slowly pushed its cultivation (and adaption to the cold) northward. Theoretically, the hardiest specimens that survived the colder winters as it moved north had genetics that adapted to colder and colder conditions and passed that along to their offspring.[2] This theory has not been proven; however, we do know close relatives of *Asimina* in the *Annonaceae* family do indeed thrive in tropical Florida, Mexico, and Central America, namely, *cherimoya*, *guanabana*, custard apples, etc. Other related edible and nonedible species of *Annonaceae* grow in many other tropical regions from Jamaica to Africa, India to Malaysia. What we do know is that pawpaws have been present in North America for millions of years, based on fossil records and their presence in pre-Columbus Native American culture. That is the history of pawpaws in a nutshell.

It is my personal theory that the primordial native forests of the US used to be much more biodiverse, awe-inspiring habitats with billions more bird denizens, such as the (now extinct) carrier pigeon, which no doubt heavily fertilized the forests via droppings and created massive food sources for vast hordes of insects, including carrion beetles and flies, which happen to also pollinate pawpaw. The carrier pigeon's habitat was huge thickets of river cane, a native plant of riparian areas which also typically is found near wild pawpaw groves. Thus, in primordial times, pawpaw patches, undisturbed and likely hundreds of years old, no doubt spanned acres of ground, slowly and gradually creeping along via underground suckers, which emerged to become new tree stems. Those same bird droppings mentioned earlier would have provided vast fertility as well as pollinator food sources, thus creating high-production and high-quality fruit stands. Whatever the case, we know that the pawpaw is currently experiencing a resurgence in its numbers due to human cultivation as well as its rebounding in the forest ecosystems of the eastern US. Thankfully, for pawpaw lovers, the US government no longer considers pawpaws a pest species and no longer encourages land owners to destroy any and all pawpaw trees, as they did in the past. Some foresters and biologists even wonder if pawpaw is not becoming somewhat invasive, or at least overly aggressive in its growth and domination of some forest under-stories.[3] This is in large part to the overpopulation of white tail deer that avoid browsing on pawpaw foliage but eat many other tree seedlings, thus giving pawpaws an advantage. Whatever the case, thankfully, pawpaw is here to stay and might be on the upswing.

Botany of Pawpaw

The pawpaw's native range is a massively large one, encompassing portions of 26 states, including most of the eastern and some of central areas of the US. Its range extends from Southern Ontario and Michigan into Indiana, Ohio, New York, Pennsylvania, and even the southern tips of near-coastal Connecticut. The whole mid-Atlantic area is pawpaw country. The entire South down to northern Florida has pawpaws (except any sub-tropical areas; remember, pawpaw is a temperate species). They also eagerly established westward to most or all of Missouri, forested areas of Kansas, and even grow wild further west, including parts of Iowa, Oklahoma, Arkansas, and the Texas Panhandle. Humans have no doubt extended the pawpaw range even further into other Midwestern areas and more recently onto the West Coast.

To make botanical matters clear, we need to distinguish between the species *Carica papaya* (also called pawpaw) and North American pawpaws, *Asimina triloba*. In many tropical countries of the former British Empire, the papaya is often affectionately called "pawpaw," and sometimes other tropical fruits are also called "pawpaw." This causes confusion when people lacking knowledge about *Asimina triloba* make erroneous statements that North American pawpaw is related to papaya, mangoes, or even bananas! This is definitely not true in the botanical sense at all. Pawpaws are in the family *Annonacae*. Papayas are in the family *Caricaceae* and bananas in the family *Musaceae*. These are not considered even closely related to *Asimina triloba* and are completely different plant families and species.

Asimina triloba is what is now known as the pawpaw tree in modern popular culture. Still, be careful when researching "pawpaw" to make sure you're actually looking into *Asimina triloba* and not papaya or some other fruit. People in some rural places apparently still refer to them as "woods banana," or the "Indiana/Michigan/Kentucky/insert-state-name-here banana," and other obscure names have been printed, such as "false banana" or "banango," but I have really only ever heard them referred to as pawpaws, which is probably best. Who wants a "false banana" anyway?

Pawpaw Myths

There are some really hilarious and sometimes off-putting folklore and myths about pawpaws still in circulation, mostly simply handed down by people who have very little or no personal experience with pawpaws. Some of these myths are so off-putting as to damage the image and desirability of pawpaw, and deserve to be immediately discarded so people stop repeating them. Here are the most common erroneous beliefs.

Credit: publicdomainpictures.net

Also referred to as "pawpaw," causing some confusion, tropical papaya is completely different from *Asimina triloba*.

Belief #1: Pawpaws are a tropical fruit related to papaya (*Carica papaya*), bananas, or mango.
Fact: Pawpaws are not tropical plants, they are a fully temperate species and are not related to any of these species, botanically speaking (although the same common name is applied to both). As mentioned already, pawpaw is in the family *Annonaceae* and is therefore related to the "custard apples," such as sugar apple (sitapol), cherimoya, and soursop (guanabana). It is distantly related to the magnolias and is in the order *Magnoliaceae*. Interestingly it resembles Southern magnolias in leaf shape, flower construction, and dormant leaf buds, if observed closely.

Belief #2: Pawpaws grow best in the forest (or in deep shade).
Fact: Wild pawpaws are usually found as understory plants growing seemingly unaffected by dense shade. However, they perform much better and produce drastically higher fruit yields when grown in full-sun conditions. They can grow and survive in either dense shade or full sun. Dense shade reduces fruit yields dramatically. Pawpaws often need sun protection the first couple of seasons, which may have led to this myth. This is all explained in great detail in later sections on cultivating pawpaw.

Belief #3: Pawpaws are only good to eat when they've been exposed to frost in autumn and/or have turned black.
Fact: The vast majority of pawpaws, both wild and cultivars, ripen weeks or even months before the first frost occurs in their native

growing range. By the time autumn frosts typically occur within their range in October/November, depending on location, most pawpaw trees are losing leaves, going dormant, and the fruits are completely gone for the season, having mostly ripened in September. In the far northern limits of their range, the fruit may ripen around the time of the first frost, but this is not common in the vast majority of their native range. Frost does not assist ripening or enhance fruit quality in any way and will actually damage the tender fruit. Pawpaws that have turned black are often still good to eat but at this stage are often considered past the prime for most people's taste preferences. How many people prefer black bananas? Frost and/or cold temperatures are also not a ripening factor whatsoever. Frost will likely damage or destroy pawpaw fruit that is still hanging in the trees, and such an event only *ever* occurs on those that have late-season ripening fruit and only in the far northern (and not optimal) range of pawpaw growing territory. People that repeat this belief are probably confusing pawpaw with persimmon (*Diospyros virginiana*), another native fruit worthy of improvement and marketing, and humorously, the frost-ripening effect is not true for them either. Persimmons can and do often fully ripen weeks before the first frost, and frost does not remove astringency in the fruit or assist ripening whatsoever. They do tend to ripen around frost time, though, explaining the misconception.

Belief #4: Pawpaw trees are immune to or unaffected by pests and diseases.
Fact: If only that were true! This misconception (sometimes lie) is mostly an oft-repeated marketing ploy that is used to sell more pawpaw trees to new uneducated growers. While *certainly* much less burdened by the plethora of diseases and insect pests that plague our apples and peaches, pawpaws are by no means immune to diseases and insect pests, and in their native range are actually susceptible to quite a number of damaging organisms. When trees are maintained and well grown on favorable sites, they are often remarkably unbothered. However, they do have their share of "plant karma" and issues to contend with, which tend to become increasingly intense and con-

centrated in monoculture plantings, as well other stressful environments pawpaws are not adapted to, or on poor soil, and generally unfavorable cultural or site conditions. This will be covered in much detail in chapter 9. That being said, most urban backyard growers will rarely experience much noticeable disease or insect pressure, and pawpaws are definitely one of the easiest fruit trees to successfully grow organically. This makes them very attractive to the niche market farmer, organic grower, or anyone looking to diversify into unusual or specialty crops. However, remember that all species of life are susceptible to diseases, insects, and parasites. Also important to note is it seems in my experience that pawpaw trees grown in rural areas in their native range are likely *much more* plagued with insects and diseases than those grown in the urban jungle, where pawpaws are in tiny isolated plantings, tucked away from their native (and other) pests.

Credit: commons.wikipedia.org

Lovely spring pawpaw flowers just shy of being in full bloom.

Belief #5: Pawpaw flowers smell like rotting flesh (or otherwise disgusting/bad/off-putting) and will stink up your entire yard.

Fact: *No, they certainly do not!* The flowers have a *very faint* unusual smell, somewhat yeasty and musky. Although pollinated by flies, the trees do not exude a horrible smell, or hardly any scent at all. The only way to even smell any aroma is to stick your nose directly up to the flower, so they will absolutely not stink up your yard, but exude an almost unnoticeable yeasty fragrance for a short time. This one has been repeated occasionally by our nursery customers and online in a very negative connotation. This one myth really has to go, and fast![4]

With the falsities now behind us, let us begin with fundamentally where pawpaws are to be found in the wild: in the deep forests and parks of eastern North America!

1

Foraging for Wild Pawpaws

We'll start off this pawpaw journey the same way I began my relationship with this most curious, elusive, and desirable of trees... searching for them in the wild. That search has not been a disappointment in any way. It has been extremely enjoyable and complex, owing to the species' very diverse and colorful past and promising future as an important multi-purpose temperate fruit tree.

Why would one want to forage for wild pawpaws? There are a number of reasons. First, it's often fun and educational, and you can see how these trees grow naturally. Second, many wild trees produce good-quality fruit that can be harvested. Third, there is much genetic potential in the wild pawpaws within their native range, and trees worthy of propagation and introduction definitely exist out there. They can also provide seed for growing trees or rootstock.

It's worth noting the eight other species of *Asimina* that grow in eastern North America. These similar but much less desirable (for fruit production) cousins of *Asimina triloba* typically do not produce desirable or even edible fruit, yet they may have characteristics to contribute to breeding experiments and are worthy of protection. Others are possible future native landscaping trees and are quite beautiful. For instance, *Asimina pygmaea*, found in South Georgia and Florida, is very dwarf[1] in stature (around 12 to 24 inches tall at maturity) but produces small fruit with little to no eating value. Breeding *Asimina pygmaea* with a large-fruited *Asimina triloba* cultivar could possibly produce a dwarf-sized tree that produces good large fruit. I came across *Asimina pygmaea* or possibly *Asimina parviflora* many years ago in a forested area near Savannah, Georgia, and thought they were *Asimina triloba*. I could not figure out why they were so small (around 2 feet tall), until later when I learned they were a different species altogether. So, we have to first make sure we are searching in the correct areas and for the correct species. There are other native subspecies that produce bigger blooms and other characteristics, but none come close to the quality and size of the delicious edible fruit produced by *Asimina triloba*. Neal Peterson has successfully hybridized different *Asimina* species with novel results (larger flowers, etc.).[2]

Unknowingly, I grew up right in the heart of the pawpaw's native range. Kentucky is known by researchers to be a prime area for locating, researching, and growing premium-quality pawpaws. Most native forested environments in Kentucky contain sizable populations of wild pawpaws. The very best habitat to find them is in moist lowland, riparian zones: the river bottom areas and muddy shores and slopes of small rivers and streams. Often they are on the sunnier east, west, and southern slopes of hills, especially in *well-draining* areas above waterways. They can also be found growing thick among the sunnier stretches of cleared trails and roads in parklands. Edges where moist meadows meet the forest are also suitable habitats. Contrary to what some people think, pawpaws do not grow in mucky, poor-draining, swampy, or wetland areas! They must have good drainage and soil aeration. You'll never find pawpaws growing in a swamp, although possibly nearby, uphill a bit on better draining land, no doubt. They cannot handle continual wet feet, meaning, having the roots immersed in water like a cypress tree. Being a species of riparian zones, they are adapted to occasionally having their roots submerged for a few days or a week with no negative repercussions. See the color photos section for a nice shot of an older wild grove in Louisville, Kentucky, growing in a riparian zone that occasionally floods.

Nowadays pawpaw trees are becoming more common in native plant gardens, urban settings, edible landscaping, and some reforestation projects. Make sure to ask permission before foraging for pawpaw fruit growing on someone else's property! Although pawpaw trees are commonly and sometimes abundantly found in natural areas within their native range, the tricky part is finding the elusive fruit. You will often find lush stands of trees but nothing more than green leaves and branches, with zero to very little fruit yield. Why is that?

To answer, pawpaw trees in the wild (and those left unchecked in your garden) do not grow into a single-trunk form akin to trees such as most pines, oaks, and maple. Their growth habit is more colonial and expansive, forming the proverbial pawpaw patch. First, pawpaw seeds inside fruit dropped in autumn overwinter in forest litter and then germinate in spring, quickly sending down a deep 12+" taproot that soon expands into a full root system. About 6 to 12 weeks later, the plant sprouts a single stem that eventually becomes a single trunk and a year later forms branches, etc., much like any tree. However, like wild plum trees (*Prunus americana*), pawpaw roots soon sprout forth many shoots, called suckers, from their horizontally growing roots, on trees between about 5 to 7 years old. These emerge from underground to eventually form a clonal stand of genetically identical tree-like suckers, all joined underground by a single root system. All of these suckers can grow to tree size (15 to 20 feet tall, 6 to 12 inches in diameter). Thus, the pawpaw tree becomes a thicket of its own that appears to be many trees closely spaced together (3 to 6 feet apart usually) but is actually only one—an original "mother tree" surrounded by a thicket of tree-size sucker "trees" of various ages and sizes. A pawpaw thicket of one specimen can sometimes span a large area, with suckers forming 10 to 20 feet or more away from the original mother tree, with dozens of tree-like shoots, creating the impression of a stand of many trees growing together. However, being that pawpaw trees are rarely self-fertile, meaning that without the nearby presence of another genetically different specimen in flower at the same time, there will be almost no cross-pollination, and thus zero to scarce fruit set. Add that to the fact that shaded trees produce few flowers each spring, that also happen to utilize an unusual pollination strategy, and you will often find yourself disappointed with the lack of fruit to be harvested from even large pawpaw stands found in the wild. Adding further uncertainty to the situation, the fruit quality in the wild varies greatly from amazing to "spitters" (fruit so bad, you just gotta spit 'em out!). In my foraging trips, I have found about 25% of the wild fruit in Kentucky to be of decent good-eating quality, 50% to be marginally edible, and the other 25% to be spitters.

You can also identify pawpaws by the bark which is very smooth, silver-gray and may often be spotted with holes left by sapsuckers. The twigs and leaves when crushed release an acrid odor I liken to the sulfurous smell of common Fourth of July fireworks.

When you do find good wild pawpaw fruit it is often *really* good and worth the effort. Foragers that take to the waterways via canoe in rich pawpaw territory tend to come back loaded with buckets of fruit and the least tired. Just keep in mind that pawpaws are only in season for about 4 to 6 weeks, from late August to late September within their main range and September to October in the northern fringe (Michigan, New York, Pennsylvania, Northern Ohio, etc.). Old-timers in Kentucky say that they used to follow the fluttering zebra swallowtail butterflies all the way to the wild pawpaw patches. Whether they are speaking poetically or factually, this might actually be useful, as will be explained later.

Wild pawpaws are easiest to identify when blooming around mid-late April. With no leaves on the trees, the purple-black bell-shaped blooms, an inch or two across, are very conspicuous against the somewhat sparse vegetation around them. Insects will also often be abuzz around the trees. No, you will not smell anything objectionable!

When you do find a good pawpaw patch, rejoice in the abundance and goodness of the pawpaw. Tread lightly, leave no trace, and do not damage the trees or attempt to climb their weak, brittle branches, as this can severely damage the trees. Shaking the branches violently will dislodge many underripe fruit that will never ripen properly. Just check underneath the trees for fruit or give them the slightest gentle shake to dislodge them. Consider helping the patch by planting some genetically different pawpaws nearby, planting pawpaw seeds, maybe cutting down a few invasive trees nearby, or removing some litter. Excellent wild specimens can be propagated via sucker removal or grafting, as will be explained in the propagation section of this book.

2

Description of North American Pawpaw Fruit

By now you know the botany and how to identify wild pawpaws. Worth mentioning is that the North American Pawpaw Growers Association has decided to start using the term "North American pawpaw" when describing *Asimina triloba*. This may be appropriate sometimes because, as explained earlier, in many tropical regions the unrelated papaya (*Carica papaya*) is also called "pawpaw," resulting in a fair amount of confusion. However, for simplicity's sake, we will assume by now you know this book is about *Asimina triloba* and not something else, and the term "pawpaw" is what we will call the fruits and trees of the species.

It can be challenging to describe what a pawpaw fruit tastes like unless you are a gourmet cook or unusual fruit enthusiast. It's fair to say "a pawpaw tastes like a pawpaw." That's true. After all, how would one effectively convey the unique flavor of a banana, watermelon, or peach? Like these, pawpaw is unique in and of itself. Most people tend to say they taste like banana, and that is a fair enough ballpark description, but we can get more much in-depth and descriptive. The fact is, many pawpaws don't taste at all of banana. Pawpaw has many possible flavor combinations, and the nuances are rich and plentiful. However, unlike most fruits, the flavor changes dramatically at the stage of soft ripe to the still edible but quite different overripe stage, with black or brown skin color.

Common positive flavor descriptions of various cultivars include: banana, mango, strawberry, vanilla, persimmon, chocolate, cocoa butter, pumpkin, anise, coconut, Mexican flan, pineapple, citrus/orange, cantaloupe, cherimoya, caramel, honey, marshmallow, nutty...that's quite a large range of variation! Few (if any!) fruit species in existence would generate such a varied and rich list of possible delicious flavors. In fact, that's downright impressive. It's possible that only Durian fruit (*Durio zibethinus*) would have such a long and varied comparable potential flavor list.

Common positive texture descriptions include: custard, marshmallow, creamy, thick, avocado-like, chewy, buttery, smooth, juicy, refreshing, moist, melting, pudding-like, soft, and silky.

Common negative flavor descriptions include: bitter skin or seeds, bland, nasty, cloyingly sweet, off-flavor, bad aftertaste, acrid, sulfurous, or just "too weird!"

The plump, ripe fruits of 'Nyomi's Delicious', a superior pawpaw from Berea, KY.

Common negative texture descriptions include: watery, slimy, pasty, goopy, gelatinous, loose, hard, rubbery, gritty, unpleasant, and gross.

In my evaluations of pawpaw cultivars compiled at the end of this book, negative textures and off-flavors of different cultivars are listed and taken seriously as strong deterrents to growers looking to grow quality pawpaws. See chapter 11, Pawpaw Cultivars for many in-depth descriptions of all known pawpaw cultivars.

A challenge with marketing pawpaw in the US is that most 21st century Americans tend to prefer crunchy or firm fruit, even now preferring crunchy peaches.[1] Mostly this preference has been dictated by decades of conditioning by commercial growers, supermarkets, and universities that have bred and marketed commercial fruit cultivars that are so hard and firm you can pick them mostly ripe off the tree, ship them 2,000 miles, hold them in storage for months, and still sell them rock-hard and unbruised, and possibly even play tennis with them. Gone are the dripping sweet with ambrosial nectar fruits of old (at least not on supermarket shelves). American tastes are accustomed to temperate fruits that tend to either be crunchy, juicy, or somewhat soft (think banana) but not *custardy*. Pawpaw is a custardy fruit, meaning the texture leans toward a soft pudding-like consistency (think whipped ripe avocado). Custardy fruits are almost entirely of the tropical sort, and rarely, if ever, found in American markets. So, when introducing people new to North American pawpaw, you should prepare them for a gustatory experience very different from apple. This has been one of the challenges of introducing pawpaws to the American customer: their exotic custardy texture seems too odd and for some, off-putting. Those texture people can be hard to please! However, I think framing pawpaw texture as similar to creamy avocado yet sweet like tropical fruit could appeal to many and prepare people for a positive introduction.

To be clear, we must understand that many, if not most, pawpaw fruits are very mediocre eating, just like some mangos or apples are mediocre eating. To have outstanding pawpaw fruit, you must have access to high-quality, genetically superior trees. Fruits also need to be very ripe, always. That being said, pawpaws that are excellent far exceed any expectations of temperate-zone fruit and can take the adventurous person to taste-bud ecstasy. Many people love them immediately from the first few bites. Many first-timers say they taste exactly like a prepared desert, such as flan or chiffon. That may be a good way to introduce or describe them to people.

Typically, two main types of pawpaws have been described in literature based on pulp color, but I am proposing there are three main varieties: yellow, orange, and white. Although there are no hard and fast rules, the orange-fleshed tend toward being stronger flavored, often with a heavier banana/honey/persimmon/pumpkin flavor profile. Yellow-fleshed, being the most common, tend toward a lighter banana/cocoa butter/Mexican flan/nutty/marshmallow/caramel, very sweet flavor, sometimes with a very pleasing tropical fruit, light citrus, pineapple, cantaloupe, or strawberry aroma. White-fleshed varieties are rare and tend to be more on the mild flavor side, with vanilla/lightly banana/cantaloupe/coconut/tropical fruit (pineapple/mango/cherimoya) and with a high sugar content. All three types can be excellent, with the yellow- and white-fleshed types likely appealing to more people. All three types are described in chapter 11, Pawpaw Cultivars. Some of the very best pawpaws I've had have been white-fleshed. Individual cultivars differ, and there are good and poor examples of all three varieties. Color can be an attractive feature in different recipes and for different purposes. For instance, white pawpaws would not turn a product yellow, whereas orange or yellow pawpaws would impart a distinct orange or yellow color.

The best pawpaws tend to have a creamy, thick, dense texture that is not especially juicy, goopy, loose, or at all watery. An excellent pawpaw, cut open and on a plate, should hold together and should not start oozing or melting out on the plate or in your hand. You can't drink excellent pawpaws; you spoon them up or lightly chew them. Some people may like goopy fruit, but it's almost guaranteed most customers would find that totally off-putting. Excellent pawpaw texture is similar to firm Greek yogurt or ripe avocado, very soft, yet thick and pleasingly creamy, with a very sweet flavor and fragrant aroma. Poor-quality fruits tend toward a loose texture like watery cottage cheese or juicy banana mush, or they simply remain overly firm and hard like an underripe avocado without ever getting pleasingly soft and edible. Excellent pawpaws have agreeable "exotic" fruity, caramel, or nutty flavor(s) and fruity or perfumed aroma, with little to no bitterness or unpleasant aftertastes. The best pawpaws also tend to have a very low seed-to-pulp ratio of less than 10% seed weight to edible pulp weight. Some modern high-quality cultivars are even as low as 3 to 6% seed/pulp weight, which is impressive. Wild pawpaws can be very high in seed weight, sometimes exceeding 50 to 75% in seed-to-pulp ratio, and this is of course not very pleasant to eat and not profitable to cultivate.

The very best ones I've eaten nearly all weighed close to one pound each, were exceedingly sweet, thick, and creamy, and contained very low seed weight. Most pawpaws contain large seeds about the size of fava beans (around 2 to 3 inches long) that are black/brown, shiny, hard, and inedible. The better-quality pawpaws tend to have smaller seeds, some as small as black beans (about 1 inch long). There are some cultivars, such as 'Sunflower', that are still considered very high quality but do contain large seeds, but despite that they have excellent texture and flavor with plenty of pulp. Some pawpaw fruits have seeds that are easily removed from the fruit; when fruits are halved lengthwise, the seeds can be removed before eating. This is a very nice feature that we'll call "freestone." Others have a tiny gelatinous sack (called an aril) firmly clinging around each seed that complicates easy removal and also limits pulp extraction when processing the fruit (yet still does not make

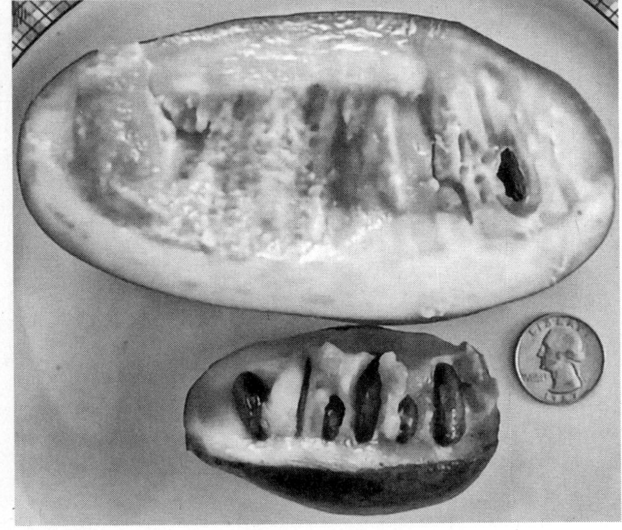

A superior cultivated pawpaw (*top*) compared to an inferior wild fruit (*bottom*). See the difference?

the fruit unpleasant to eat). We can call this "clingstone." Some growers say the age and ripeness of the fruit might factor into the removability of seeds and presence of the arils. Others claim refrigeration of the fruit makes the seeds easier to remove, but I have not found this to be the case. Cultivar differences are also at play. Some cultivars, like 'Susquehanna', are always freestone and never have arils. Both freestone and clingstone pawpaws could be considered good or excellent to eat. Some report that the aril may contain bitter flavors, depending on the cultivar, so it is probably best to avoid mixing them into the pulp when processing pawpaws, if you are able to.

Complaints About Some Cultivars

Of course, pawpaws *would* have some weird quirks to them, including the potential of strange aftertastes and even possible nausea from eating them. One common complaint with some cultivars (even some considered good cultivars) is a lingering bitter or slightly unpleasant aftertaste. I have not found a slight bitter aftertaste to be off-putting, but some people are more sensitive to this. Recently, most newer pawpaw cultivars considered to be excellent have been selected to be free of bitter aftertaste. However, if some pawpaw cultivars are said to have strange or poor aftertaste, they are best avoided in order to prevent customer rejection. 'Sunflower' cultivar pawpaws are sometimes accused of having a slightly bitter aftertaste, but otherwise are still considered very good-excellent.

Nausea from consuming pawpaws is a rare but real concern. A very small percentage of people simply cannot tolerate eating the fruit, and it might make them feel mildly sick or nauseous, even to the point of vomiting! Two different men have told me they love pawpaws but cannot eat them because they are allergic to pawpaw skin/rind. In my early foraging days, when I naively picked and ate firm, slightly underripe fruit, I experienced mild nausea afterward, as did the other person I was foraging with who also ate a lot of them. It's best to never eat, process, or offer someone underripe pawpaws! If selling the fruits in the firm–ripe stage, which is usually best because they are more durable, let customers know they usually require ripening at room temperature, like an avocado or green banana, for a day or two until slightly soft and fragrant. Let people know they are not edible until soft. Some only like pawpaws when they are "dead ripe" to the point of being black-brown in skin color, similar to overly ripe bananas.

If a customer, family member, or friend eats pawpaws and experiences nausea, they likely won't eat (or buy) them ever again, so make sure they're good and ripe; if someone starts to feel queasy, don't give them more pawpaws! It's important to note here that some people cannot tolerate *cooked* pawpaw, even though they love fresh pawpaws. I am apparently one of those people. My wife used to make us pawpaw bread and even this thick, silky, luscious pawpaw custard pie, which we relished, until a few pies later, even the smell of it cooking started to make us both feel queasy. Scientists have not identified the cause behind that. Some people think mixing and eating pawpaw with grains (such as a wheat crust on a pie) can lead to nausea. So, unfortunately some people might experience this negative side effect as well. I think it's best not to mention it because any pawpaw side effects are rare, not dangerous (just unpleasant), and bringing it up might create a placebo effect in people who are anxious, food-neurotic, or skeptical about pawpaws and other exotic foods in general. If someone gets nauseous because of eating pawpaws, they'll realize it on their own.

Now, what about the pawpaw "buzz/euphoria"? My goal is to cover all aspects of pawpaws, even strange or controversial ones. Well, this is not mentioned much, but some folks do actually claim that consuming pawpaws can make one feel a little on the "very mildly intoxicated" side very temporarily. A local man and fellow pawpaw enjoyer, as well as my wife and I, have experienced this independently numerous times, and sometimes eating pawpaws makes me feel sleepy and/or extremely relaxed, which I find very pleasant. This very brief side effect seems to pose no danger or problem, and most people might find this very appealing, or not experience it at all. There is no scientific data on why this may occur. It's probably best not mentioned to customers as it may either scare them away or disappoint them if it does not occur. Most people do not experience or do not notice any such mild side effects, either positive or negative. However, I wouldn't recommend a pawpaw feast on your first date with a special someone. These uncommon side effects are all just potential aspects of the pawpaw experience. For the vast majority of people eating pawpaws, they'll only experience eating a delicious fruit and have no surprise side effects whatsoever.

A warning: Never make, sell, or eat pawpaw fruit leather! Pawpaw fruit leather is a toxic product and causes temporary yet serious gastrointestinal distress. In case you do not know, fruit leather is a term for mashing fruit, spreading it very thin, and dehydrating it. This thin, translucent, leathery fruit product, similar to packaged fruit snacks for kids, is wonderful when made from berries but toxic when made from pawpaw. Neal Peterson says 5 or 6 people became nauseated from pawpaw fruit leather in his research. If consumed, it makes most people very nauseous and ill for approximately 24 hours.[2] Symptoms include nausea, vomiting, and diarrhea. Not fun! Don't do it. The reason is unknown but could be bacterial contamination, or oxidation of compounds such as fatty acids going rancid during the drying process.[3]

Acetogenins

Finally, what are acetogenins? In a nutshell, these are not very well understood, yet very bioactive, fatty-acid compounds found to varying degrees in pawpaw fruit, leaves, and twigs and in other species of *Annonaceae*. Some pawpaw cultivars are high in acetogenins, some low in them. KSU is currently breeding fruit to be low in acetogenins and also breeding for fruit high in acetogenins.[4] It's not totally clear if these compounds are beneficial or potentially harmful. There is evidence of anti-tumor and anti-cancer activity[5] yet also evidence of potential links of acetogenins to Parkinson's disease.[6] However, the likelihood of developing health problems by eating a few dozen pawpaws for a month every year seems very unlikely, as any tests done showing possible links to adverse health effects were done researching consumption of a different species of acetogenin-containing *Annonacae* fruit (guanabana/soursop)[7] and is based on regular, ongoing daily consumption. I wouldn't worry much about it. Just make sure you're not eating pawpaws or pawpaw products every day, or gorging on huge amounts of them. That being said, it is my opinion that pawpaw breeders and researchers should be breeding and selecting fruit low in acetogenins in case more substantial evidence comes out that these are potentially harmful to health.

Physiological Characteristics of Pawpaw

Botanically speaking, the pawpaw itself is a hypogeal, dicotyledonous, shrub-like tree of the family *Annonacae* in the order *Magnoliaceae*. Its native habitat is the understory and stream/river banks of moist, humid, temperate deciduous hardwood forests of eastern North America. The tree is typically extremely cold hardy to –20°F, which collaborates with USDA Hardiness Zone 5. It produces the largest edible tree fruit in North America, and is technically a berry.[8] Pawpaw fruits can weigh over 1.5 pounds (24 ounces), but very rarely exceed or reach 2 pounds (32 ounces). More breeding work could create trees that regularly produce 2- to 3-pound fruits (yet this could prove to be overwhelmingly large for most US customers). Typical pawpaw fruits weigh from 3 to 12 ounces; those in the 7- to 10+-ounce range are considered suitable for marketing.

Pawpaw fruits are born in clusters, attached to a single peduncle,[9] composed of 1 to 8 individual fruits (rarely more), usually 2 to 6. When mature, these clusters resemble hands of bananas, all radially emerging in a spherical shape from the single peduncle attached to the branch. The peduncle is brown, short (around ½" to 1"), slightly fuzzy, and attaches firmly from the twig of the branch to a bulbous stem base that is attached to the top of each fruit. The amount of fruit per peduncle depends on the success of the flower pollination and genetics of the cultivar. Some cultivars tend toward producing single fruits per peduncle (and less effective pollination also can lead to one fruit per peduncle). One fruit per peduncle is now considered

A multi-clustered pawpaw here with 6 individual fruits on 1 peduncle. Delicious and impressive, yet not ideal for marketing due to the tear in the flesh to separate them, which will quickly start to ferment within 1 to 2 days.

a positive trait for marketing purposes, as each fruit does not need to be torn off separately at the peduncle (which prevents exposing the flesh to the open air where it can begin to rapidly rot). This will be elaborated on in the chapter on pawpaw harvesting.

Pawpaws are small trees, sometimes considered more a large shrub than a small tree. Maturing to around 20 feet tall and 18 to 20 feet wide in full sun, they are definitely a small tree when compared to oaks and maples. However, the ultimate height and shape of a pawpaw tree varies greatly depending on the amount of light the tree receives. When growing in the shady forest understory, where pawpaws are usually found, the tree takes on a reaching, stretched, and limber form, with large branch angles, larger leaves stretched out more parallel with the forest floor, reaching 30 to 40 feet tall after many years. In such a forested shady situation, pawpaw usually produces relatively few flowers and fruits, and the trees are of a much taller, open form. However, when cultivated in an open area in full sun or if a chance wild tree grows in a more sunny location, the tree takes on a much more densely branched, squat, pyramidal form with more acute and tightly aligned branches, shorter stature, and slightly smaller, darker leaves that droop protectively down to the ground, making for a very attractive specimen. When grown in full sun, the amount of flowers produced and the potential yields of fruit are much, much higher.

The exotic attractive flowers resemble velvety, crimson-purple orchids or banana flower petals. They are considered "perfect" flowers, meaning they contain both male pollen-producing anthers and female pollen-receiving pistils. They appear first as swollen, black fuzzy buds on the trees in autumn the size of a BB. The nonflowering leaf buds appear as tiny flat black-brown fuzzy crescent-shaped buds along the stems and twigs. In early spring, usually in mid-March or early April, the flower buds quickly enlarge, swell, and soon open up into mature flowers. The rubbery, blackish-purple bell-shaped flowers hang down, protecting the delicate pollen-studded anthers and awaiting sticky pistils from the frequent and very heavy spring rains, effectively shedding the rain and allowing pollination to continue un-

Credit: Cliff England

A top pollinator of pawpaw is the common fly. Attracted by the faint yeasty odor of the abundant flowers and the protein-rich pollen, the fly soon finds itself covered in pollen and moving it from tree to tree.

The attractive, smooth silver-grey bark of pawpaw is recognizable all year round. It never turns patchy or scaly with age. Native Americans utilized the tough, stringy inner bark to make very strong rope.

abated. The flowers are pollinated by a number of odd forest denizens and not by honeybees or bumble bees. Pollination will be covered in more detail later. Flies, fruit flies, carrion beetles, and other shady creatures attracted to the musky, lightly acrid scent of the flowers are solely responsible for moving about the pollen from ripe anthers on one flower to awaiting sticky stigmas on another. Pawpaw flowers are non-self-pollinating due to the fact that the anther's pollen is usually ripe only after the pistil is receptive. This is common in many plants and prevents self-pollination or, in other words, inbreeding. Humorously, some people deride the pawpaw pollinators as being "ineffective" at their job, even though they have obviously succeeded in very effectively pollinating *Asimina triloba* as long as the trees have existed, probably millions of years.

3

Flowering and Pollination

For full detailed color photos illustrating each stage of bloom, please see the color section of the book.

Flowering

The pawpaw primarily flowers and fruits on 1- and 2-year-old wood. Fruiting buds are formed in late summer and do not bloom until the following spring, becoming next year's fruit crop. So, unlike fruits such as grapes, persimmons, and mulberries that bloom and fruit on new season's fresh green growth, pawpaws only flower on wood produced the year or two before and *never* on new growth produced that same season. In fact, they flower early in the season before vegetative or leaf growth has even commenced, and the tree appears bare and dormant still, except for the little dark-purple bell-like flowers dangling gracefully upon it (a sure sign spring has arrived). The pawpaw does *not* form fruiting spurs (tiny specialized flowering branches) like apples and pears. The flowers are born singly just above the axils (leaf scars) of the current (or last) season's leaves. Flowers have 2 whorls and 3 petals, and the calyx has 3 sepals. Each flower has multiple ovaries, which is why multiple pawpaw fruits can be born in clusters on the tree. However, only strong pollination produces multiple fruit clusters, and that is not always desirable for marketing purposes.

Stages of Flower Bloom

Like any fruit tree, or any flowering tree for that matter, the process of flowering can be broken down into stages of development, from tiny dormant flower bud to fully opened and enlarged, viable and fertile flower to petal drop and fruit initiation. With apples and pears, for instance, the process of flowering is divided into distinct stages by horticulturists and orchardists. The reason for this categorization is to understand what the flowers' vulnerability is to factors such as frost and freezing weather, as well as spray concerns and pollination needs. With this knowledge comes the opportunity to understand potential threats and attempt to protect the flowering trees from harm. Common threats include freezes or fungal spores and are dealt with by providing frost protection or some sort or fungicide spray applications, respectively. Frost protection usually assumes the form of either very enormous industrial electric fans or a water spray misting system, both of which can reduce or eliminate frost damage. The viability of such protection systems in pawpaw cultivation is so far not documented. A third method, although less reliable but worth noting here, is the application of kelp extracts (*Ascophyllum nodosum* seaweed species) which, according to many reliable sources, will increase cold hardiness by 2° to 4°F, if applied as a foliar application 6 to 12 hours before the cold temperature event. Another useful potential protection is the use of old-fashioned orchard heating lamps powered by kerosene, propane, or diesel.

Note that pawpaw trees have a very long bloom period that can last 3 to 4 weeks (longer if you count the entire process from smallest start to finish, explained in detail ahead). Pawpaws usually have several, if not all, stages of bloom happening simultaneously. In other words, a tree can have flowers in the Velvet Bud stage all the way to Full Bloom and beyond at the same time. This is excellent because if a late hard freeze event occurs it will sometimes only damage or destroy the blooms in the furthest along stages and thus the remaining, and less developed (but more cold-hardy) buds will continue to develop and fruit. This makes pawpaws hardy and resilient, and most years, they will fruit even when a late freeze occurs (that would wipe out an entire crop of, say, peaches or apples that tend to bloom all at once).

I am attempting here to define and categorize the stages of pawpaw bloom for the purpose of creating a common language with which to communicate the stages of flowering to help with practical purposes surrounding pollination. Any knowledgeable apple growers, for instance, understand immediately what is meant by "green tip" and "pink" when discussing the stages of apple flowering, and they understand the limits of cold hardiness and other vulnerabilities of the trees at that time. It seems wise to have a similar classification system for pawpaw flowering. So, with this purpose in mind, here are the stages of pawpaw flowering as I have observed and named them:

Dormant: The flowering buds are completely dormant, tiny, and hardly noticeable. They are contained inside a protective hard casing and appear as bronze-brown tiny spheres just above the dormant vegetative buds on the twigs. They are about the size of the "O" on this page. At this time, the buds are traditionally assumed to be cold hardy to at least –20°F on most cultivars. Some southern US cultivars could possibly be less cold hardy, but all seems cold hardy to at least –10°F, and probably colder.

Velvet Bud: The flower buds now appear as tiny black, fuzzy, spheres located just above the dormant vegetative buds on the twigs and are about the size of the "O" on this page. The buds are in the process of shedding, or have shed, the protective casing over them (which appears as tiny cup-like plates that open in halves). This usually occurs around mid-February in Kentucky, USDA Hardiness Zone 6. Exact cold hardiness is not precisely known but is generally hardy to and maybe slightly below –20°F.

Velvet Stem: The flower bud has now enlarged and extended itself from the twig. It now has its own distinct stem that is clearly seen. The flower bud stem is thick, black, and covered in velvet-like fuzz. At this point the bud is around the size of a small pea and appears to be covered in black fuzzy hairs. This usually occurs in early March in Kentucky, USDA Hardiness Zone 6. Fuzzy stem buds have been observed on our farm to be cold hardy to at least 18°F with no damage whatsoever. So, they can freeze and/or get snowed and frosted on and apparently do not get damaged.

Green: The bud now enlarges even more, turns from fuzzy and black to green and appears lightly fuzzy like a peach. Green buds are about the size of a large pea and have differentiated from a sphere shape and now appears to have 4 or 5 distinct, fused together sepals. This usually occurs in late March or early April in Kentucky, USDA Hardiness Zone 6. Cold hardiness is not determined, but a light frost (not a freeze) will likely not damage the buds at this stage. A freeze will usually destroy the buds at this stage.

Pink: The bud now has enlarged to about marble size, the sepals have begun to open slightly, and distinct green petals can be clearly seen, which often but not always have a pinkish-purple blush to them. It is possible but not confirmed that some buds at Pink stage may be viable to receive pollen, as stigmas will appear sticky and elongated inside the unraveling flower. If ripe pollen from another genetically different tree is also available at this time, it would be advisable to begin hand-pollinating buds at the Pink stage now and/or encouraging pollinators. This stage usually occurs in early April to early May in Kentucky, USDA Hardiness Zone 6. Cold hardiness is decreased but has not been determined. A light frost could likely be tolerated still. A freeze will usually destroy the buds at this stage.

Full Bloom: The flowers are in full glory of bloom with petals fully extended outwards, sloped, somewhat tongue shaped; they appear a distinct dark maroon-brown or against bright light will appear bright reddish. Stigmas are definitely ripe: shiny, sticky, and thus ready to receive pollen. Anthers are fully formed, tightly packed behind the stigmas, white, and not viable. Full Bloom flowers are about the diameter of a quarter (2" to 3"), hang downward to shed rainwater and should be abuzz with pollinators and various insect species. This stage usually begins in early April to early May in Kentucky, USDA Hardiness Zone 6. Flowers remain in this stage with viable stigmas for approximately 48 to 72 hours.

Frost and temperatures at and below 32°F will destroy flowers at this stage. If frost or freeze is predicted, preventative measures should be taken at this time, such as misting systems, orchard heaters, fires, covering small trees with row cover fabric, covering very small trees with barrels, or Christmas lights.

Pollen Shed: This is the final stage of pawpaw bloom and would be easy to overlook. The petals now begin to slightly wilt, dry up, and/or shed. The stigmas are now no longer viable and cannot receive pollen. The anthers behind the stigmas are now extended and covered in ripe fluffy, whitish-yellow pollen. If pollination was successful via pollen from a genetically different tree, the stigmas at this time will often appear swollen, bright green, and beginning to extend slightly from the base. Thus, Pollen Shed is also the first stage of fruit development.

The pollen readily comes off the anthers and clings to the bodies of pollinating insects, or a paintbrush or other tool used for hand-pollination. Insects flock the flowers at this time to feed on the nutritious pawpaw pollen. This stage usually begins in late March to mid-May in Kentucky, USDA Hardiness Zone 6. It lasts approximately 48 to 72 hours or less, depending on when the pollen supply is extinguished. Windy or rainy weather will remove the pollen from the anthers; if hand-pollination is the plan, do so in dry conditions or before a rain. Pollen Shed can last about 3 or 4 weeks as the tree gradually goes through its annual flowering cycle.

At this point, the young fruitlets appear as tiny slightly fuzzy green "fingers" (up to 9 or 10 or more) or single tiny fruitlets, dangling from the now rapidly enlarging peduncle fruit stem (formerly the flower stem). Hanging downward, each is adorned with a tiny black dot on the end of each fruitlet, resembling tiny banana fruits.

Utilizing a variety of early-, mid- and late-ripening cultivars can potentially stretch the harvest period from July or August into October. Large-scale (future) commercial growers might appreciate cultivars whose flowering cycle is shorter; thus, fruit ripens in a smaller window of time, saving harvest labor and costs. However, most small-scale growers and marketers will find this long season advantageous for securing a good crop as well as providing a long marketing window of time in which to sell the fruit and gain the interest of customers. Because pawpaw does not keep more than a few weeks refrigerated, this is also advantageous.

Pollination

It's necessary to discuss the curious matter of pawpaw pollination. As with most aspects having to do with the nature of the pawpaw, nothing is quite conventional or simple. As explained earlier, the pawpaw is not pollinated by bees and does not appear to be effectively wind-pollinated (like oaks or maples). Rather, it is insect pollinated, but by flies (*Diptera* sp.), including houseflies, green bottle flies (blow flies), and various species of fruit flies; beetles (*Nitidulidae*), including possibly carrion beetles; and other forest denizens attracted to the musky, acrid scent of the flowers.

I have observed ladybugs, various spiders, ants, and what appeared to be Asian longhorned beetles each covered in, and feeding on, pawpaw pollen amidst the flowers in full bloom. Missouri State University adds bumble flower beetles to the list (*Euphoria inda*).[1]

After observing and studying the many different species of insects involved in pawpaw pollination, I have determined the following:

- Pawpaws appear to be an important and abundant source of protein-rich pollen very early in the spring that attracts many diverse insect species.

Known Pawpaw Pollinators

- Flies, including houseflies, green bottle flies (blow flies), and fruit flies
- Ladybugs (*Anoplophora glabripennis*)
- Click beetles (*Elateridae*)
- Ants (various)
- Spiders (various)
- Bumble flower beetles (*Euphoria inda*)

Possible Pawpaw Pollinators

- Asian longhorned beetles (*Anoplophora glabripennis*)
- Carrion beetles

- There are probably dozens of insect species capable and active in pollinating pawpaw trees that have not been identified yet.
- The best means for naturally pollinating your trees is to encourage and protect biodiversity in your orchard setting.

Conventional thought is that flies are generally the best pollinators of pawpaws, and this appears correct simply because they are definitely attracted to pawpaw blooms and they are also highly active and extremely mobile. So, definitely attract flies when the pawpaw grove is in bloom. This can be done any number of creative ways. One is to simply place items in the pawpaw patch that strongly attract flies: hang ripe bananas and peels in the trees or place roadkill, dead fish, strips of bacon, spoiled meat, dead squirrels or opossums, fresh manure, etc. around the orchard. Sounds great, right? Some people tie little bundles of meat or even commercial catfish bait from the branches. Replace these once a week with a fresh supply. Strangely and somewhat disturbing is my observation that cheap commercial frozen sausages and lunchmeat attracted very few to no flies (I thought fly bait might be its only suitable use, but I was wrong; it apparently has no good use whatsoever). Soon thereafter, the now attracted flies will also check out the musky flowers of the nearby pawpaw trees, thus getting covered in pollen and bringing that to ripe stigmas, pollinating the flowers. And, folks, let's not talk about this practice with potential pawpaw customers or news reporters, ever. It's enough to turn some people off from pawpaw fruit before they've even tasted one and could reduce sales and create bad public image. The fact is, if you have an orchard full of blooming pawpaws, the pollinators will find them. Yields may be less, however, but that depends on other factors, such as local environment and species diversity.

Instead of focusing solely on flies, the better strategy is attracting and encouraging as many insects as possible by maintaining flowering undergrowth in the rows between trees, wherever there is not mulch. Planting (or at least not mowing down when flowering) plenty of early-season flowering trees, flowers, and herbs—such as comfrey, black locust, lilacs, flowering bulbs, dandelion, clovers, and fleabane—will attract plenty of insect life capable of pollinating pawpaws. And, by all means, attract plenty of flies.

The motley crew of insect life just described are the ones solely responsible for moving the pollen from ripe anthers on one flower to awaiting sticky stigmas on another, and hopefully between genetically different trees! This means the flowers are not self-compatible nor wind-pollinated (like corn, grass, walnut trees, etc.). However, in some rare cases, single pawpaw trees can exhibit some amount of *self-pollination*, which means one tree can pollinate itself and bear fruit. However, this fruit is usually lower quality and in small quantity compared to a tree that has been well-pollinated by another genetically different tree. Never rely on current claims of self-fertility with pawpaws. This could change with breeding efforts perhaps.

You can also hand-pollinate the trees. This is the best choice for those seeking optimum yields, if the labor is available. This is also a good strategy when overly wide tree spacing or isolated trees would hamper natural pollination or when pawpaw crossbreeding is desired for creating new cultivars.

Hand-pollination

In order to hand-pollinate pawpaws, timing is essential, careful handling and lots of patience is in order. Hand-pollination should be done on a low-wind and dry day. Pluck and collect flowers from trees that are currently in the Pollen Shed stage. You'll notice the wilted petals and, upon close examination, the shaggy white-yellow pollen inside the flower. The flowers are sometimes still blooming but just barely starting to wilt or brown at this stage. Carefully pluck them off the tree and store them in a little container, like a plastic food storage box. When ready to pollinate another tree, peel back the petals of a Pollen Shed flower and, with a small soft-bristled artist's paintbrush (a camel hair brush is often recommended), remove pollen from the stamens, and then, using the pollen inoculated paintbrush, dabble it lavishly onto the awaiting sticky stigma of *another tree's* ripe flower in Full Bloom or Pink. Make sure the flower is at the stage of pollen-reception, wherein the tiny stigmas will appear sticky or shiny. By the time the Pollen Shed stage has arrived, that flower will not be receptive to fertilization via pollen. Make sure that the Pollen Shedding flowers being used are from a different cultivar or otherwise genetically different tree than the one you're pollinating! Otherwise no fruit set will result. Differing and compatible genetics would be found between any two seedlings, a seedling and a named cultivar, or two different named cultivars ('Sunflower' and 'Overleese', for instance).

Hand-pollination, when well performed, is documented to often result in very heavy fruit production. For instance, in 2017, I hand-pollinated a younger but mature size 'Sunflower' tree and counted over 100 resultant fruits on it. In 2018, I did not hand-pollinate, and the same tree bore only about 20 to 30 fruits, even though other flowering pawpaw trees were present.

Oftentimes hand-pollination is *so* effective it can cause the trees to over-produce. This can cause two issues: excessive weight on the branches and smaller fruit. Thinning will often then be necessary to maximize fruit size and quality and potentially prevent branches from snapping under heavy fruit loads! These are mostly the concern of commercial growers and not backyard producers. This will be covered in more detail later in this chapter.

The basic thing to understand is that, in the Southeastern US, natural insect pollination is usually sufficient to produce good yields of excellent fruits. Pawpaw trees should be clustered in groves or set in rows no further than 10 to 18 feet apart for proper pollination. Wide spacing (20 feet or more) will certainly lessen pollination prospects and can reduce or even eliminate fruit set. In this case, hand-pollination would be the only immediate option. Planting more pawpaw trees (in order to close in the gaps between the existing trees) would be the other more effective long-term option.

What about "self-pollinating" pawpaw trees? Like we mentioned earlier, sometimes an isolated pawpaw tree has managed to produce some fruit. Self-pollinating pawpaw trees are possible, but appear to be more a survival mechanism of the tree itself; in years when it receives no cross-pollination, it can somehow produce some amount of inbred fruit and thus seeds. Some varieties may perhaps be more prone to this behavior than others. The fact is, though, self-pollinated fruit will be low quality, and one should never rely on any claims of self-pollinating pawpaw varieties and always have at least two genetically different pawpaw trees present in any orchard or planting. As far as pawpaw pollination is concerned, the more cultivars or seedlings present, the better. A commercial planting should definitely have a minimum of 3 or 4 cultivars present and even some quality seedlings to assure proper cross-pollination.

Thinning of Fruits

Should you thin the fruits? Perhaps. It depends on the fruit load, the destined final use and market for the fruit, as well as your available time and labor resources.[2] If they are destined for local markets as premium-quality (and premium-priced) local fruits, you will want to thin them as best you can to achieve large size and excellent quality. If you are growing them for processing or for home use, thinning is not necessary.

Thinning pawpaw fruits is still an experimental technique, so the exact details will develop in time. Do not attempt to thin the fruits too early in development because they are very fragile at that time and entire peduncles can accidentally be snapped off. The best time to thin is likely when the fruits are highly differentiated and individualized within the cluster and are each several inches long. This will likely be in late May or early-mid June. Do not wait too long to thin the fruit or the tree's energy and resources will have mostly already been spent, and you will simply be reducing the final yield. This is similar in practice to thinning apple fruits, which is done at nickel-size. Use a very sharp pair of hand pruners or hand snips and cleanly cut off the smaller fruits, carefully avoiding cutting into, scratching, or damaging the single fruit you intend to save. Hand pruners with dull points, like those used for thinning grapes, should work well. As with all fruit thinning, always thin to the biggest, best-formed, and healthiest-looking specimen, removing any deformed, smaller, damaged, or otherwise inferior fruits. It's also essential to thin any fruits that look likely to get sunburned due to sun exposure.

Kentucky State University thins fruit at about one-half inch in length prior to the June drop, with good results. They noticed yields by weight were the same, but the fruits were larger and better formed, with only one fruit per peduncle. June drop is a natural tendency for a tree to drop a portion of fruits every June, shortly after pollination is completed. It's a way the tree manages its own resources and does not overbear.

The reasons for thinning the fruits are several. First, thinned fruits will result in larger and higher-quality (yet less numerous) fruits. Second, when strong or hand-pollination results in multiple fruit clusters, the individual fruits ripen at different times, although still within the same peduncle cluster. For instance, a cluster of three fruits on one stem may each ripen consecutively 2 to 3 days apart. However, when a fruit within that cluster ripens, it often falls from the peduncle, having ripped off the peduncle. In other words, all the fruits within a cluster share one peduncle stem; when a clustered fruit falls or is picked, it tears off at the stem end. That stem-end tear somewhat seals over, but not completely. This small tear in the skin greatly reduces storage and shipping life of the fruits, becoming an open invitation for bacterial and fungal invasion and fermentation. So, fruits grown and harvested in this manner keep only 2 to 3 days at room temperature and about 2 weeks under refrigeration (if refrigerated the day of harvest). However, fruits that developed singly as one per peduncle (either through thinning or poor pollination or perhaps varietal characteristics) will each have their own stem (and will not need to be torn from the peduncle). When harvested with the stem still attached, the fruit keeps much better, has a better appearance and will not have an open wound that causes rapid deterioration of the fruit. However, fruits that are intended strictly for processing need not be thinned, as size and appearance do not matter as much and fruits will likely (and should) be utilized very quickly after ripening. Any fruits intended for premium direct-market sales, shipping, or storing at room temperature more than a day or two should be thinned in the grove to a single fruit per cluster whenever possible and harvested with stem attached.

Fruiting and Ripening

By June, the fruitlets should be 1 to 3 inches long or so. They continue to rapidly grow all summer. By the end of June, good cultivars should be about lemon-sized. The young fruitlets take on a rubbery, lightly bloomed (powdery) appearance that is highly resistant to rain, insects, fungus, and disease.

Most early cultivars, such as 'Nyomi's Delicious', ripen usually by mid-August in Kentucky; by the end of August, mid-season cultivars should begin ripening in Kentucky. Northern growing regions will be 1 to 3 weeks later than this. However, most pawpaws ripen from mid-August to early October, with the majority (mid-season cultivars) ripening in September.

Wild pawpaws ripen between late August to late September. Some years, the weather patterns may bring on a slightly early or later harvest, but September is really pawpaw time. Growers much north of Kentucky should make sure to focus on planting earlier-ripening cultivars, usually bred or found thriving in northern areas. The cultivar 'NC-1', from Ontario, is an excellent example, as are 'Pennsylvania Golden', 'Halvin', and 'Nyomi's Delicious'. Otherwise, you may find your pawpaw trees getting frosted on before fruits are fully ripe and losing the crop every year. Paying due diligence and choosing cultivars carefully for your region will make or break your farm operation or grove. See chapter 11, Pawpaw Cultivars.

Ripening of the fruit can be challenging to discern. Some varieties show a light-yellow coloration (color break) when ripe. 'Nyomi's Delicious', 'Golden Moon', and 'Pennsylvania Golden' are cultivars that usually show a color break to yellow, especially 'Golden Moon'. Others simply start falling from the trees, a sign the tree is ready to harvest. When marketing fresh fruit, you will not want all the fruits falling from the trees because rocks, stubs of weeds, and other things can easily bruise or damage fallen fruits, rendering them unmarketable (or considered contaminated). A very thick layer of straw mulch or thick tarps or sheets under the entire canopy should lessen drop damage.

Don't shake the trees to dislodge fruits; shaking rips the fruit off the peduncle, thus exposing fruit flesh to the open air and allowing rot to set in quickly, and also dislodges many underripe fruits, which almost never ripen properly. They will remain hard and inedible or only become slightly edible and not very palatable. Fruits ripe enough to fall will not keep very well unless refrigerated immediately, unless the stem is still attached. Once the peduncle is torn from the fruit, it will begin decomposing rapidly and usually only lasts a week or so in refrigeration and may not be marketable.

A good sign of ripeness is when the fruit is slightly soft and gives to a light press, like a peach. If this leaves a finger indention, the fruit is certainly ready to pick. A fruity aroma should also be present, and likely the tree will have started dropping a few fruits already. The fruit will also usually fade to a light bluish-green or slightly green-yellow, but not always. As explained earlier, some pawpaws show a color break at ripening, but most do not. The fruit can then be eaten or left to sit for 1 to 3 days at room temperature to reach an even more ripened state, signaled by the darkening of the skin to a brownish-yellow, like a banana. Ripe fruit tears open very easily with a light twist or a knife, the flesh is soft and usually custardy, and the flavor should be excellent and sweet. As mentioned before, underripe fruit tastes unpleasant and can cause nausea and mild gastrointestinal discomfort. Also, some people claim sensitivity and allergic reaction to the underripe fruit skin, similar to (unrelated) mango allergies, including mild rashes on hands and mouth.

Ripe fruits will keep at room temperature for 2 to 3 days and if immediately refrigerated can be stored 1 to 2 weeks in good condition, sometimes up to a month. As noted before, fruits that are born singly through thinning or weak pollination and which have the peduncle still attached at harvest will keep and ship much better and resist fermenting and molding.

4

Site Design and Planting

One thing to remember with pawpaws is that how they grow in natural settings is actually not conducive to cultivating them for high-yield fruit production. In the wild, pawpaws are almost always an understory tree, found among and under thick, dark, climax, or secondary tree growth, usually near water, forest edges, small forest clearings, or trails. However, as discussed in chapter 1, fruit harvests in the wild can often be little to nothing due to several factors.

Site Selection

Pawpaws need excellent site conditions in order to thrive. Mediocre or poor orchard sites will greatly diminish success with pawpaws. Start by selecting a site fully exposed to excellent, strong direct sunlight.

Sunlight

One of the main factors to success is available sunlight. I often give people the easy phrase for growing pawpaws (as well as most fruit species), "the more sun, the more fruit." The more direct sunlight the trees have access to once they are in their "adolescent" stage, the faster they grow and the more compact and pyramidal shape they ultimately assume.

All day, or 12 to 14 hours, direct summer sun, gives maximum results and will facilitate healthy, dense pawpaw trees with strong fruit production. An orchard site with 7 to 8 hours of strong undiluted direct sun will be adequate if there is no other option. The less direct sun, the more openly branching and lanky the trees will become, with substantially less flowering and fruit production. However, if your site is shady or wooded, pawpaws can still succeed and become a source of food and/or a minor forest product, although fruiting and production will be much reduced. For that reason, shaded and/or forested environments are not a viable option to growing pawpaws commercially.

Soil

Pawpaw trees are flexible when it comes to exact soil conditions but are not adaptable to everything. They can do well in heavy clay, sandy soil, silt soil, loam, and rocky sites. However, they require soil that is very well-draining in order to thrive and perform well. Soils that are mucky, swampy, or wetland are unsuitable, as are those that hold rainwater for days, puddle excessively, or have very high water tables. Ideal soils would be deep, loamy, well-drained, and slightly acidic to neutral. A pH of 5 to 7 is ideal. Deep soils that are 4 to 6+ feet deep before hitting bedrock or water are ideal. Shallower soils can still work but should be at least 3 feet deep.

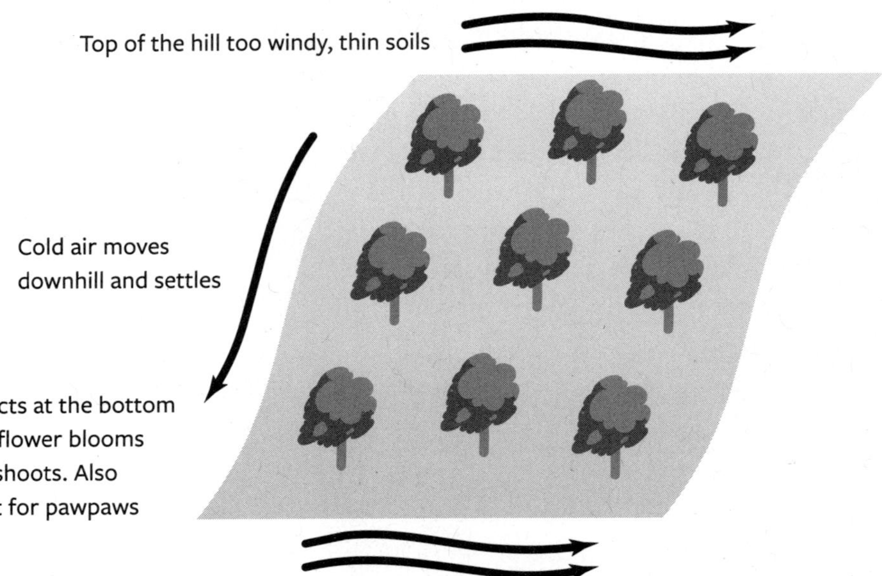

High organic matter content will foster healthy growth. This can often be ascertained via soil testing. Organic matter is crucial to healthy soil because it holds moisture, creates air spaces for oxygen, slowly releases micronutrients, and feeds the soil food web of microorganisms, earthworms, and fungi. Remember the environment where pawpaws originate: deep, rich, alluvial soils with extremely high organic matter content, slightly acidic, and usually near water. Try to recreate this the best you can, while providing full sun.

Your local Agricultural Extension Office or Soil Conservation Office will be very glad to help you understand what soil type and conditions are on your site or a new site you are considering planting or purchasing. Before purchasing or planting on any site, you should consider it an absolute necessity to consult with them first and also get a soil test. Don't risk it! Research and explore a new site carefully before investing time, resources, and energy into planting on it.

Clay

Clay soils can grow very healthy pawpaw trees as long as they are well-draining and have plenty of mulch, organic matter, and/or compost on top and you pay careful attention to preparing the planting hole. When planting on clay soil, dig a medium-sized hole, at least 12" by 12". Overly large holes can hold water and drown out the tree, so be careful. Fracture the sides and bottom with your spade or shovel. Using your hands, spade, and/or a rototiller or tilling device, aim to get the soil removed from the hole as crumbly as possible. You don't want heavy clods and blocks of soil in the hole. If the soil is very thick and hard to break up, add a small amount of compost and try again. Get the particles as small and broken up as you can (do this only in dry soil conditions!). Then very carefully place your tree roots or root ball into the hole and slowly refill it, avoiding placing large clods of soil back into the hole or dropping them onto the sensitive roots. Remove any stones you find. Gently tap down the soil. Now water the tree in and cover with thick mulch. Only plant the tree as deep as it was growing in the pot, and don't bury the trunk.

Sand

Sandy soils can be successful if lots of organic matter is added as a mulch layer, and added nutrition is given in the form of organic fertilizers, compost, and manures (always on top of the soil). Thick mulch will help hold in moisture and keep the soil cool. Sandy soils tend to dry out rapidly if not mulched. Pawpaw roots cannot tolerate very dry conditions.

Trees grown in sandy soil will most likely need regular irrigation with drip tape or drip emitters on orchard tubing so the roots do not dry out. Basins, or water wells, made around the tree will help capture moisture when watering or during rains. Basically, with sandy soil you simply need lots of mulch, a little extra fertilizer or manure, and you must keep an eye on soil moisture and water when needed.

Loam

Loam is simply a mixture of sand and clay soil types. There are sandy loams, clay loams, and silt loams. All well-draining loams should be excellent for growing pawpaws. Silt loam is perhaps the ideal soil as it resembles best the pawpaw's native soil that is often alluvial silt based. Depending on your rainfall and soil moisture, irrigation might be needed for optimum results. Heavy mulch is always required. Make sure to choose a site with well-draining soil that is not overly wet or mucky.

Rocky Soil

Rocky soil is potentially decent pawpaw ground, and it usually has excellent drainage. However, the soil must be deep. For instance, soil that seems rich and crumbly on top but a foot or two down hits shale or limestone will be poor ground for growing pawpaws and should be avoided. Be careful of hilltops and ridges as they almost always have thin soils and experience strong winds, both of which are not good for pawpaws. If planting in lowland soil or hills that are rocky, dig extra big holes, remove as many rocks as possible, and backfill with garden soil or soil scraped from nearby. Top-dress the trees with plenty of compost and mulch. Well-draining rocky soils will benefit from basins around the trees.

Wet Soils

Wet, heavy soils are not suitable for growing pawpaws. Soil that seems unusually heavy when digging, holds water for days after a rain, or has holes that fill up with water when dug are all unsuitable. Soils that smell nasty and anaerobic, and when dug display grey or bluish coloration, or currently host reeds, rushes, cattails, or crawdads are not suitable and will stunt or kill pawpaw trees. Choose another site. However, if that site is your only option, then consider talking with your local Soil Conservation District about tiling to improve drainage. You could also plant the trees in very large raised mounds above the usual level of the soil. Unfortunately, all these methods will still yield less than optimal results and will considerably increase labor and resources needed to establish a planting.

Other Site Considerations

Wind

Windy sites are to be avoided. High winds damage and tear pawpaw's large delicate leaves, dislodge flowers and fruits, and disturb pollinating insects. Extreme windchill can also damage buds, twigs, and branches. Even in central Kentucky, USDA Hardiness Zone 6 (minimum temperature –10°F), we experienced a few days of record-breaking windchill down to –30°F in the winter of 2015. This is cold enough to possibly damage pawpaw trees. Our trees were on a somewhat sheltered site and did not appear to have any damage.

A windy or very open site will need windbreaks installed in order for your pawpaws to perform optimally. Designing and planting windbreaks is beyond the scope of this book but can be found easily from other sources. Eastern white pine and poplars make excellent windbreaks in the eastern US. A rule of thumb to remember with windbreaks is that they typically slow down the wind speed and force for about three times the height of the windbreak. So, a 25-foot-tall windbreak of, say, white pines would be protecting and area up to 75 feet away downwind from the trees.

Ideally a site should be open and cleared of other trees, but surrounded by tall wind-breaking trees. It should be cleared on the downwind side, enough to allow cold air to flow down the slope. Avoid planting in frost pockets (low-lying areas). Remember, frosty cold air sinks, goes downhill, and settles in the low spots, where frost becomes concentrated. Such areas can drastically reduce fruit yields through frost damage to spring blooms.

Cover Crops

If working with clay, sand, or relatively poor (but well-draining) soil, cover cropping for one or two seasons is a good practice if you have the patience. Clover, ryegrass, sorghum-sudangrass, vetch, and Austrian winter pea all make suitable green manure cover crops. Once removed or worked into the soil, they produce a more crumbly and aerated soil with higher organic matter and nitrogen content.

Irrigation

Proximity to irrigation water is very important. Once established, pawpaws can handle average hot, sometimes low-precipitation summer conditions experienced in most of the eastern US. However, a hot, very dry spell (drought) during *tree establishment* can harm or kill an entire planting. Neal Peterson tells the story that, in the early 2000s, he planted an entire orchard of his patented pawpaw trees on a high West Virginia hilltop. That first summer, an intense drought took place. His pawpaw orchard had zero irrigation setup, and the drought conditions on the hilltop killed the entire planting, and he lost the farm. Don't let this happen to you!

In other regions with drier conditions, including Texas, Oklahoma, Arkansas, the Southwest, and California, make sure the site has irrigation water access. This could be a retention pond, a large rainwater catchment tank, or city water. For more detailed information about orchard irrigation, see chapter 6 and the books in the Recommended Reading section.

Access

Another aspect of site selection is making sure the planting has good access for vehicles. A market orchard needs to have trucks and/or tractors access the site for mowing, bushhogging, applying organic sprays, and hauling harvested fruit. Depending on how you plan on managing these details, make sure that paths and tree rows are wide enough to facilitate mowing and maintenance and allow vehicles or machinery throughout the orchard. Some people use four-wheelers for the bulk of these tasks, except mowing.

A well-designed and thoughtfully planned planting is much more likely to result in a productive, successful planting. Before acquiring trees, first calculate how many pawpaw trees you will need. Well-grown pawpaws of good genetics are quite productive; one mature (8- to 10+-year-old) tree can produce 20 to 40 pounds (or more) of marketable fruit per season. So, 100 healthy trees could potentially produce around 2,000 to 4,000 pounds. That's quite a lot of pawpaws! See chapter 12 for more information about deciding how many trees to plant.

Designing the Planting

After selecting a reasonably conducive orchard site, as detailed earlier, you need to design the planting. Conventionally, pawpaws are planted similar to other fruit trees like apples and pears. A grid of straight rows with 10 to 12 feet between trees and 18 to 20 feet between rows works fine. However, there are other designs that work well and are much more efficient.

Unless you plan on mowing the grass between rows with a tractor, they need not be 18 to 20 feet apart. If you plan on controlling this grass through grazing animals, scything, string trimming, or some other means, they can be closer; 15 to 18 feet would suffice in that situation. Be careful not to plant the rows too close together. Once the trees get large, you do not want them shading each other out, which can reduce yield.

Generally, pawpaws need close spacing to facilitate strong pollination prospects, so 8 to 12 feet between trees within the row is adequate.[1] They can also be grown in groves planted on a 10- to 15-foot center, the distance between each tree in every direction, the most efficient way of spacing. Imagine a honeycomb shape, where a tree is in the middle of each cell, 10 to 15 feet apart. They can also be planted bordering woods or other plantings, in single long rows, on contours, and in blocks.[2] They make very attractive edible landscaping trees around the home or other buildings. The only necessity is other genetically different pawpaw trees present for pollination, preferably a minimum of 3 or 4 different trees planted closely together. Also, it should be noted that pawpaws can be planted in clumps of 3 to 5 trees only a foot or so apart, but this would be for backyard only and diminishes yields.

Planting the Trees

Planting pawpaw trees is not much different than planting any other tree (at last, pawpaws aren't trying to be different!). Except, alas, they do have a few unique quirks. Remember that their delicate root system is susceptible to transplant shock, and once shocked, pawpaws don't put on much branch (top) growth for that season. A shocked tree may only grow 3 to 4 inches the first year, or may not grow at all, or may perish. A non-shocked tree in excellent conditions will typically grow 6 inches to 3 feet the first year, depending on age, size, site, cultivar, etc. The first year in the ground, realistically, 6 to 18 inches is considered good growth. Oftentimes, the trees will put down root growth the first year or two and little top growth. As slow as this first-year growth sounds, healthy pawpaw trees require up to two years to establish strong deep roots. After year 2 or 3, top growth speeds up dramatically; healthy trees should be 7 to 10 feet tall by 5 to 6 years old, and fruit production should be happening.

Potted Versus Bare-root

Most pawpaw growers, including me, would agree that (well-grown) *potted* pawpaw trees survive and establish much better and easier than *bare-root* trees. I know of only two legitimate retail nurseries that sell and ship bare-root pawpaw trees. I have planted dormant bare-root pawpaw trees, in spring, extremely carefully, and had perfect results. However, it's much safer and better to go with good potted trees as pawpaws have such a delicate and fragile root system, and I would not trust most nurseries to handle bare-root pawpaws properly.

Tree Planting Supplies

Before planting, you should have acquired the following materials. For more information and sources, see Resources in the back of the book.

- A sharp spade or shovel (or several if there will be several planters). Spades have mostly flat blades that come to a straight, flat edge (unlike the curved, spoon-like scoop of a shovel) and are much better for digging planting holes. Spades are somewhat hard to find these days, but specialty horticulture supply stores sell them and quality vintage ones are often found at flea markets.
- Kelp extract diluted in water and available onsite (plain water works fine too, but diluted kelp extract provides nutrition, minerals, and beneficial plant hormones).
- Rock phosphate. 1 pound per tree in the planting hole (optional but recommended). May be difficult to find.
- Mycorrhizal root inoculant slurry (optional but recommended).
- Mulch materials (required). Have these piled onsite, ready to use.
- Irrigation tubing. Have this waiting onsite before planting, ready to install.
- Tree shelter (required). Install these immediately after planting each tree.

Let's examine a few of these things. *Kelp extract* contains trace minerals and plant hormones that will help any plant to better handle transplant shock. It also provides crucial nutrition your trees need (micronutrients and trace minerals). Watering in each newly planted tree with 2 to 4 gallons of water or, preferably, diluted kelp extract gives optimal results. Follow directions on your kelp extract product for dilution rates. Buying powdered kelp extract as opposed to pre-mixed extract diluted in water is much cheaper. Consider kelp like a multi-vitamin/mineral for your trees.

Rock phosphate is a natural rock mineral high in phosphorous (P) and trace minerals. It is usually labeled as 0-3-0 or 0-4-0 (the middle number being the percentage of phosphorous in the product). Because these are crushed rocks, it takes 8 to 10 years or so to fully decompose and depends on soil microorganisms helping in that process in order to be effective. At tree planting, apply 1 pound (about 1½ cups) into each planting hole, mix with the native soil, and then plant the tree. It provides a long-term source of phosphorous, the essential nutrient that facilitates strong and healthy root-flower-fruit production. Nitrogen is easy to source, phosphorous not so much, so consider using this product. It is around $0.75 cents to $1 per pound, typically. Bone meal is a quicker releasing and easier to find product, but does not last as long, and is a slaughterhouse by-product.

Mycorrhizal inoculant is recommended. Only purchase high-quality mycorrhizal products from companies that specialize in these, test the fungal spore count, and provide such information. These inoculants contain arborhial fungal spores that quickly colonize plant roots and help facilitate a healthy root system that utilizes nutrients more effectively. To apply, sprinkle a teaspoon of inoculant powder on the root ball or roots at planting. Some people make a slurry-like paste and coat the roots that way. Neal Peterson says most soils should contain these naturally. Soils treated heavily with chemicals likely do not.

These products are not crucial but why not go for optimum results? Any organic or chemical-free growers would benefit from these products, as would growers using chemicals.

Flower Bloom Stages: *Left*: Formed during autumn, these tiny, BB-sized little furry buds are the first stage of flower buds, hardy to about –10°F to –20°F. They will stay this size until March or early April depending on location. *Middle*: The second stage of flower bloom occurs in later winter or early spring. The bud has now enlarged significantly, has a visible stem protruding from the woody twig, and has turned green. It is very cold-hardy, handling subfreezing temperatures without injury. Sepals tightly closed. *Right*: Sepals now begin to separate as the bud continues to rapidly expand. Still hardy to subfreezing temperatures.

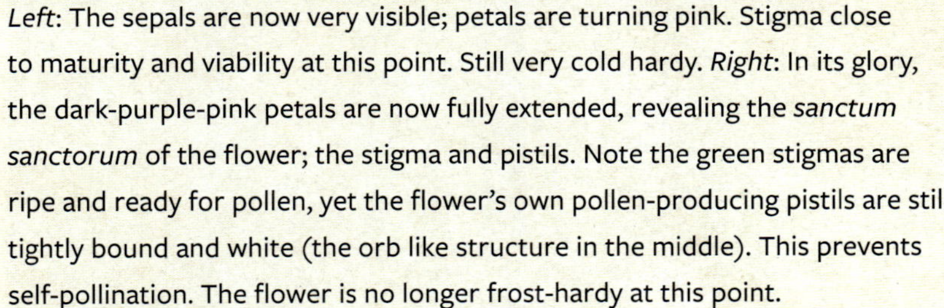

Left: The sepals are now very visible; petals are turning pink. Stigma close to maturity and viability at this point. Still very cold hardy. *Right*: In its glory, the dark-purple-pink petals are now fully extended, revealing the *sanctum sanctorum* of the flower; the stigma and pistils. Note the green stigmas are ripe and ready for pollen, yet the flower's own pollen-producing pistils are still tightly bound and white (the orb like structure in the middle). This prevents self-pollination. The flower is no longer frost-hardy at this point.

Here two stages of bloom are present, ripe pollen (*right*) and greenbud (*left*). Note ripe white pollen grains inside flower on right. Photo taken April 10 at author's grove in Louisville, KY.

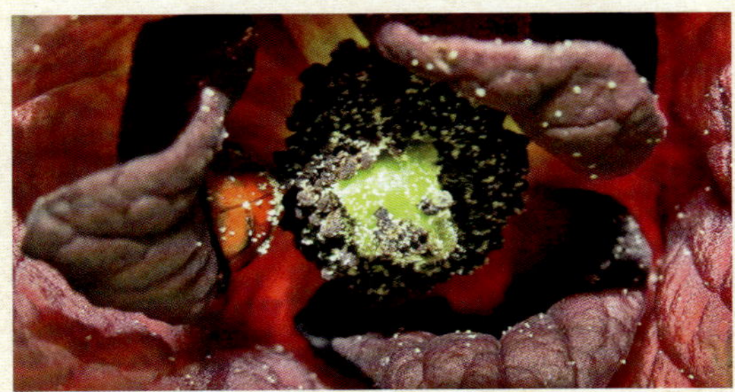

In this final stage, the pollen shed stage, the stigmas are no longer viable and not able to receive pollen, and the pistils are now full of ripe pollen. See the whitish pollen grains? This is what you collect for hand-pollination. The petals will often start to brown or appear wilted at this stage.

Left: Now post-bloom, the successfully pollinated pawpaw flower shows tiny fused-together fruitlets at the beginning stages. These sometimes will fall off. Any unpollinated flowers will drop soon as well. These are very delicate. *Middle*: About 4 to 7 days after successful pollination, the tiny fruitlets no longer are fused together and start to differentiate. Strong pollination will show multi-fruit set, although the final number of mature fruits may be less. Note banana-like appearance. *Right*: About 1-month post pollination, the fruits on improved cultivars should be around lime sized, rubbery, and bright green-blue. If you have not thinned fruits and wish to do so, now (or better yet, even a little earlier) is the time. The rubbery skin is highly resistant to fungus and insects.

Some cultivars naturally tend toward single fruitlets, such as 'Nyomi's Delicious'. Pictured is a single fruitlet on 'Mango' cultivar. This is better for premium fruit sales because the peduncle can be left attached to the fruit, not torn off at harvest like a multi-cluster fruit bunch necessitates. Photo taken in early May in KY.

Behold, the mature pawpaw! Nearly ripe for the picking, these look great and show the large size and better appearance of a highly selected superior cultivar. Photo taken in mid-September in KY.

Along Beargrass Creek in Louisville, KY, there are a few wild pawpaw groves that show how primordial pawpaw habitats likely appeared. This area is about 8 feet above a large creek, and the trees grow in the muddy, rich alluvial soil that briefly floods every few years.

The lovely and talented pawpaw expert Rachel Cothron stands in front of a large 20+-year-old pawpaw seedling grown in 100% full sun in a field at the Lexington Kentucky UK Arboretum. Notice the fully filled out, circular canopy made by dozens of suckers, as well as the central main trunk. A seedling tree in this condition can apparently live 75 to 100 years, maybe more.

Lush, healthy, vibrant pawpaw growth in mid-spring. Pawpaw shoots should rapidly grow by 18 to 24 inches the first 5 to 7 years, or you may need to apply more fertilizer and/or compost.

A young pawpaw in its second season in the ground from a small grafted start. Note large healthy foliage, menorah-like shape, and tube protection. Also note the two top branches that are forming a competitive U-shaped form. This will need to be corrected by removing one branch or spreading one to a 60° angle.

Author's pawpaw planting (one of three) established in 2017. Note Blue-X tree tubes on every young pawpaw, and very thick grass mulch cut and applied in place. This grove established very well, and by the end of 2019, many trees were 3 feet to over 5 feet tall.

'KSU Benson' sapling in the author's grove, showing strong growth one season from grafting. Note the chicken wire cage for protection and the mulch. Typically. I remove all growth below about 24 to 30 inches, leaving only the main shoot to push upward and eventually start branching. You don't really want branches starting 6 inches up the trunk.

Without careful observation and maintenance, a root sucker can quickly overpower the grafted tree. Large suckers can damage the trunk when removed.

A young "adolescent" pawpaw showing strong menorah-like form and healthy green leaves and shoots, growing on north side of home in rich fertile soil. This specimen is only about 2 years from a small grafted start.

Premium-quality pawpaws like these ripe 'Susquehanna' speak for themselves, yes? This is what you want to have to offer your customers, but it takes some time, patience, labor, and resources to acquire. Price accordingly.

Zebra swallowtail butterfly (ZSB) larvae at several stages. These could all be present at the same time as well as unhatched eggs. The top largest larva is near maturity. The larva grows rapidly and should be monitored in a nursery setting or brand-new planting.

Pawpaw leafroller at work. Note webbing which typically rolls up the leaf like a burrito, safekeeping for the tender hungry larva inside.

Trogus penator wasp devouring ZSB larva on pawpaw tree. These hungry wasps will help control a multitude of caterpillars, if allowed space to exist. Fear not!

Like many webworm species, these little guys appear almost overnight, make a thick sticky protective web, and quickly start defoliating entire twigs and branches. Easily removed on lower branches, these require observation to notice their mid-summer presence in the orchard.

Top: This severe *Phyllosticta* infection has damaged this fruit and rendered it mostly worthless. Note thick black tissue damage and cracks. These cracks invite microorganisms and yeasts to move in and start doing their thing, and they don't delay. *Bottom*: Moderate *Phyllosticta* infection on ripe fruit. Note the unsightly black blotching and brown scarring. Luckily this fruit is still edible, likely unaffected, and even possibly marketable (or at least able to be processed) due to some amount of inherent genetic resistance to this fungal disease.

Phyllosticta first shows itself on leaves, then maturing fruits. Note grey-brown necrotic spots without any colored halo. This fungus eats holes in the leaves and makes black splotches on the fruits.

Japanese beetles have found pawpaws to be a suitable food source, but damage is usually not bad. Watch out for them on young saplings, though, starting around late June.

A problematic fungal disease of pawpaw, blackspot creates sickly black spots with yellow halos; the leaves can also turn yellow. This disease is not well understood yet and can kill entire trees. It has been observed in the author's grove and also on wild trees.

Credit: Ron Powell

Only, so far, a serious threat in the Pacific Northwest, and not well understood, Blue Stem Disease (BSD) attacks the vascular system, turning it a sickly blue upon inspection, killing the host tree. Note the light blue-stained cambium bark.

Credit: Blake Cothron

Mostly a problem in greenhouse situations, thrips and whiteflies can colonize pawpaw trees. Their feeding on leaves and sap will suck leaves dry, making them appear bleached, whitish, and brittle. Apply essential oil sprays labeled for insects and make sure trees are fertilized and getting strong airflow via fans.

Credit: Jerry Lehman

This tiny beetle attacks in spring out of nowhere and can kill or severely damage young and even older trees alike by burrowing into the trunk. Note the toothpick-like frass (mixture of excrement and sawdust) that is the sure sign Asian ambrosia beetles are present.

Not a disease, this is a nutrient deficiency caused by trace-mineral deficiency of magnesium and/or potash. After adding Epsom salts to the tree, it did not reoccur the following season.

Three superior pawpaws. *Top*: 'Prima 1216'. *Middle*: 'Taytwo'. *Bottom*: KSU selection, unreleased. Note freestone seeds.

Three more superior pawpaws. *Top*: Jerry Lehman selection '275-27'. *Middle*: 'KSU 7-1' is a superior pawpaw with great flavor and texture. *Bottom*: 'Susquehanna' fruit, the favorite release of Neal Peterson, showing its huge size and superior appearance. Flavor and texture are outstanding.

Planting

When planting pawpaws (or any fruit tree), digging the planting hole properly is crucial. A poorly dug hole can badly hamper a planting, or even ruin it. Good tree holes are properly deep, wide, and have *fractured* sides and bottoms so new roots can easily penetrate, not hitting a polished-like sidewall or a rock-hard bottom. It's not recommended to use an auger on the back of a tractor, but should you do so, absolutely take the time, if you are physically able, to dig out the hole a bit more and fracture the sides and bottom. Holes should be at least 18 inches wide and deep. Bigger is even better in well-drained soils. In very heavy clay soils, it might be better to dig the holes less deeply, so they don't become waterlogged. However, practicalities are central; if you have lots of trees to plant, try to make each hole at least 18 inches by 18 inches. Holes should be circular in shape and not square or blocky. Just remember to fracture the sides so roots can penetrate outward easily and do not start circling the hole. If you simply rip a hole open with an auger and throw the tree in the hole, it's not going to turn out as good, and failure may result. Pawpaws are delicate and need extra attention at planting.

Let's set down a few basic rules for planting pawpaws:

1. Only plant pawpaws in spring or early summer! It is believed that a pawpaw tree's root system goes dormant in autumn and does not continue to grow from about October to March or so. This is unlike apples and pears, whose roots continue growing all winter, and thus autumn planting is an excellent choice. Pawpaw trees will often suffer a few broken or snapped roots at planting. If planted in spring this is no problem because the trees will soon resume growing, heal the broken roots, and quickly expand their root system. However, if pawpaws are transplanted in autumn, the dormant trees will not heal over the broken roots and the root system will thus rot and the trees will likely perish. Some growers advocate planting trees only in early spring just as buds are beginning to swell. This gives excellent results. Also, if done carefully, potted

trees can be very carefully planted later in the season after having leafed out, but will experience minor shock and need extra water to establish well. It's not recommended to plant pawpaws in the heat of summer.

2. Transplant on cool cloudy days and not hot sunny, windy days. The less shock on the root system, the better. Never let dry winds hit exposed root systems. Remember, even a few hours of hot sun exposure can burn young leafed-out pawpaw trees, so keep the them in the shade of a truck, under cover or plant early in the morning or late in the afternoon, and shelter each tree immediately before moving on to the next.
3. Make sure to add 1 pound of rock phosphate per hole, and coat roots with a fresh mycorrhizal plant dip or powder, if possible. It's worth purchasing and using these special ingredients.
4. Plant pawpaws at the same depth they were growing in their pot. Do not plant deeper and certainly not shallower.
5. Be as careful as possible when placing the root ball in the hole. Planting trees grown in deep pots when the root systems are not very filled out can disturb the roots because the root ball will fall apart when removed from the pot. In that case, very carefully take a sharp knife or razor and cut out the bottom of the pot about 3 inches from the bottom, making an incision all around the pot, and remove the bottom section completely. Then, make a vertical cut all the way from the top of the pot down to the (now removed) bottom of the pot. Do not remove the pot yet. Place the tree (with the cut pot still holding the soil together) in the center of the hole, pile up about 10 to 12 inches of soil around the pot, and then carefully slide the cut pot up and out of the hole, with one hand holding the tree in place and your other hand sliding the pot upward and off. Refill the hole slowly and carefully with very loose soil, firming it along the way, very gently. In this way, the soil stays intact and the roots do not collapse into the hole.
6. Give plenty of water and/or diluted kelp extract. About 2 to 3 gallons is sufficient, pouring slowly so as not to erode the hole and expose roots.

7. If planting bare rootstock, which is not recommended, do so extremely carefully. Gently spread the roots out, refill the hole with very loose crumbly soil, spreading the roots in all directions. Refill the hole to about 1 to 2 inches above where the topmost roots are formed. Don't plant too deeply or too shallowly to avoid exposing roots.
8. After filling in the planting hole, very gently tamp the soil with a light press of the feet. Do not pack the soil in firmly or stomp the hole! Packing the soil close to the stem can damage roots—not good.
9. Apply plenty of mulch material and tree shelter.

Tree Protection

Please realize that tree shelters are a necessity, not an option. All seedling pawpaw trees less than about 30 inches tall and any grafted pawpaw trees less than 18 inches will require protection from direct UV light at planting. Tree death usually occurs rapidly due to burning, without protection from the strong direct UV radiation from the sun. Many nurseries do a great disservice to their customers by not informing them of this reality! Entire plantings can and do fail for this simple, completely avoidable reason.

Tree shelters take several shapes and forms, many of which are available on the market and all of them probably work fine. However, the main necessities are that they are least 30 inches tall, sturdy, staked securely, and block UV light. Tubex and Blue-X tree shelters all work well but need to be staked either by small wooden or rebar poles or bamboo/river cane. These need to be about 5 to 6 feet tall so that they can go underground and also extend at least 1 foot above the tree shelter tube. Be careful not to stab the stake into the ground too close to the pawpaw tree and thus damage its root system. Look inside the tube and make sure the little tree is growing straight in there and correct it if it's bent downward. Likely the tree will appear awfully crowded in there, and being in the tube may bend and deform the leaves, but it will still grow well. I have witnessed, and had our nursery customers tell me, their foot-tall pawpaws grew out the top of the 30-inch tube the very first season in the ground. At that point, the tube can be removed.

We have found the very best method comprises 18-inch wide tubes fashioned out of 30-inch high chicken wire, topped with greenhouse shade cloth (several layers of cheesecloth also works), and secured with 5 or 6 wooden clothespins or wire. Secure the tubes to the ground with a rebar post, the tube being tied together with wire or woven through the post. After a season or two of sun protection, the trees should be reaching the top of the chicken wire tube, at which point the shade cloth is removed so as to not hinder growth.

Anything sturdy that provides shade will work, but you could buy a box of 30-inch or taller tree tubes if you have hundreds of trees to plant. Just remember: please don't think they're not necessary! Your entire planting will probably fail if you don't utilize UV shade/tree protection! These tubes also help make the trees grow straight and prevent damage from deer, string trimmers, and wind damage.

The main downside of plastic tree tube shelters is that they hold humidity. High humidity and lack of air exchange are suitable environments for the proliferation of fungus. You may notice that your trees show symptoms of *Phyllosticta* while growing in plastic tree tubes. See chapter 9 for detailed information. The symptoms of *Phyllosticta* are basically a season-long black blotching on the fruit and leaves, which sometimes leads to some amount of defoliation and, in severe infections, causes fruit to crack and split. Obviously, this is not good, but we have observed that the trees still grow in the tubes, even if they show these symptoms, and the symptoms should lessen or disappear once tubes are removed. Because chicken wire tubes offer excellent air exchange and ventilation, this infection is generally not seen. This is the only method we utilize now. Also noteworthy is that wasps will sometimes make nests in plastic tree tubes.

Here's some important advice: some growers have found that the best practice is to remove tree tubes in September, after a full season of using them in the field, because they create a warmer microclimate that delays hardening-off of the new green shoots into winter-hardy wood. When severe cold hits, this immature green tissue is damaged or destroyed, which can even lead to tree loss.[3]

After removing all plastic tree protection tubes in September, you can install a hardware cloth or chicken wire protective tube as permanent protection from voles, mice, string trimmers, etc.

Comparison of Tree Shelter Systems

Blue-X Grow Tubes

Pros: Cheap, easy, and fast to install. Effectively prevent sun scald and tree death. Prevents zebra swallowtail butterfly (ZSB) infestation by excluding ZSB (see chapter 9 for full details on ZSB activities and prevention). Protects trees from string trimmers somewhat. These tubes force the tree to grow straight yet sometimes weak due to lack of wind stress. Make sure to use 30-inch tubes, not 24-inch or any other size. Must be used with stakes (5 to 6 feet of bamboo or river cane works well and lasts one season). Must be removed in September to allow proper hardening-off of plant tissues.
Cons: High-humidity conditions within the tubes can lead to *Phyllosticta* infections and sometimes (but usually not) leads to tree demise. Would work better if ventilation holes were cut in, perhaps with a paper hole punch. These are fragile and last only 1 to 2 seasons. Must be removed in September (October down South).
Cost: About $1.50 to $2 each

Tubex Shelter

Pros: Rigid, more durable, easy, and fast to install. These offer more room for growth. Prevents string trimmer damage effectively.
Cons: Bird landings will deposit tree weed seeds onto the pawpaw, which will sprout and must be removed. Fairly expensive. Wasps nest in them unless netted on top. May lead to *Phyllosticta* issues. Must be removed in September.
Cost: About $2 to $4 each

Chicken Wire Cages with Shade Cloth

Pros: Most durable, effective, and long-lasting option. Can be left on indefinitely and/or saved and reused indefinitely. Prevents string trimmer damage and vole/mice and other animal damage. Gives excellent unimpeded airflow so *Phyllosticta* is not encouraged to infect the tree foliage. Can be left on for years, year-round. Shade cloth is removed once the tree is 30 inches tall. Overall gives the best results, hands down. Use a rebar stake or metal electric fence post to hold the chicken wire in place.

Cons: Most expensive option and takes the most time to setup and install. Chicken wire must be hand cut and wired together. Shade cloth or cheesecloth and wooden clothespins are another small expense, calculated below.

Cost: About $2.50 to $4 per tree

5-Gallon Buckets with Bottom Removed

Pros: This method certainly works and is practical. They protect from UV light, animals, and string trimmers. Five-gallon buckets can be tricky to obtain, but delis and restaurants sometimes will give them away. Avoid using ones that previously contained harmful chemicals.

Cons: Expensive if bought new, tricky to obtain for free, especially in quantity. Some people may not like the look of them. May not provide adequate UV light protection, but can be covered in shade cloth, cheesecloth, etc.

Cost: Can be obtained free; new are about $4 to $5 each

It should be noted that shade-tolerant pawpaws can be interplanted in-row in between larger trees or trees that will eventually shade the pawpaws. This includes Chinese chestnuts, black walnuts, pears, etc. If those trees are planted 20 ft apart, pawpaws can be tucked in right between (20 ft between pawpaws, then). This will give you an additional, perhaps faster, crop (pawpaw fruit) and can be a big dra w to a U-Pick or diversified orchard or silvopasture. Select, improved seedlings would work great for U-Pick.

5

Choosing Your Trees

Now that you have your site determined and are sure it is suitable for pawpaw culture, as well as a nice planting design on paper, you need to know exactly what type (cultivar) of trees to plant, and how and when to source high-quality healthy specimens. With that ascertained, you may want to begin acquiring your trees, as it could take 6 to 12 months or longer to procure high-quality trees from reliable, quality nurseries. I cannot stress enough how crucial it is to source only the highest-quality pawpaw trees. This chapter will help you decide how and why to choose what to grow: quality seedlings, grafted trees, or a combination.

The first step in planting any fruit tree is selecting desirable cultivars from a reliable source. Pawpaws, as of 2019, can be obtained in many different places and forms: potted, bare-root, retail, wholesale, grafted, seedlings, dormant and leafless, or fully leafed out, 6 inches or 6 feet tall, etc. There are at least 20 different cultivars widely available (many more in collections and trial plantings), but which ones are the best? Knowing how and where to choose the right trees for your situation is very critical to success.

Before we go into acquiring planting stock, let's burst the bubble right now around planting 5- to 6-foot tall pawpaw trees "in order to quickly create a productive pawpaw orchard." Large trees transplant very poorly and are extremely expensive.

A large pawpaw root system is crucial to the tree's success and establishment, and until the tree establishes its roots, growth will be very moderate to slow. Planting a smaller 2- to 4-foot tall pawpaw tree with good roots is virtually always better than planting a 5- to 6-foot tree in a pot. Don't be tempted to overspend on big trees; they won't actually get you ahead unless they have huge root systems, which most nurseries are not providing. You'll end up spending a lot more, and the large trees will likely experience much more stress in the field and not succeed. Trust me, it will be a huge waste, and you'll have to start over after pulling out lots of 5- to 6-foot tall dead or stunted trees. Many scam nurseries online try to hock these at high prices. Instead, look for healthy, properly grown, well-rooted, well-nursed pawpaw trees of small to moderate size (at least 1 foot, preferably 2 to 4 feet) and forget the "bigger is better" mentality, which is usually false, especially when planting fruit trees. All that's bigger is the expense and also the risk. Focus on excellent nursery stock, an excellent planting site, and excellent care (mulch, fertilizer, weed control, tree protection, irrigation), have patience, and the results will be good. If all you can locate is very small nursery trees (5 to 12 inches), then either keep shopping

Unhealthy Growth
Causes: poor nursery stock, poor soil, weed/grass competition, unhealthy root system, lack of water and/or nutrients, or damaged tree. Needs replacement.

1st Year 2nd Year 3rd Year 4th Year

Healthy Growth
Note strong branching, trunk girth increase, and distinct, menorah-like branching pattern.

1st Year 2nd Year 3rd Year

around or consider nursing them in pots for a season till they get larger and more resilient. This is challenging, though, so the best option is to buy high-quality, medium-sized potted stock only.

Cultivar Selection

Cultivar is a horticultural term that describes a specific tree or plant that has been selected, named, and further propagated by people. For instance, 'Golden Delicious' or 'Fuji' are both cultivars of apple, *Malus domestica*. Cultivars are desirable to plant because they give you uniform, predictable, and profitable results, when done wisely. An orchard of 'Fuji' apples will yield bushels of mostly identical 'Fuji' apples (not random quality apples like a seedling would produce), which are in high demand by name and fetch a good price. A wise apple grower would never gamble or guess on what cultivar of apple to grow; this could, and likely would, ruin their operation. This section will explain the cultivar selection process. For very detailed descriptions of pawpaw cultivars available, see chapter 11. Take note that the word *cultivar* is incorrectly yet often (and usually) synonymous with the word *variety*.

Deciding which cultivars to grow can make or break your orchard operation. An incorrect or poor choice is a problem not remedied easily, and the time and effort to attempt to remedy it could cause major financial or marketing difficulties, so this is a crucial step in your planning.

Credit: Rachel Cothron

The author holds a grafted pawpaw trees on 3-year-old rootstock, just grafted about 2 months prior to taking this photo. High-quality trees 2 to 3 feet tall are the optimal size for planting out an orchard.

Planting your site entirely in seedling pawpaw trees is not recommended because fruits may not be premium quality. Those looking to process pulp as the main operation may choose to buy seedlings, but make sure they have an excellent genetic background, generally found through planting seedlings grown from seed of excellent cultivars.

Start the cultivar selection process by determining your main marketing niche and route. Fresh fruit sales or processing? Do you want large, plump, impressive-looking fruits to sell for a premium price? When you have that decided, read over the list of cultivars in chapter 11 and be as practical as possible in finalizing your cultivar decisions. Do not guess, do not be random, and do not rely on nurseries to steer you in the right direction or give you accurate information. Many of the sales reps and clerks you might speak with have never eaten pawpaws and know virtually nothing about them, even though they sell the trees. At best, usually they will repeat brief descriptions based on trying to make the trees sound good to get sales (e.g., large fruit, very productive, excellent flavor). This may or may not be accurate information, so be cautious.

If you live in the northern range of pawpaw cultivation, then only plant cultivars known to perform well in northern regions ('PA Golden' series, 'NC-1', 'Halvin', etc.). Many common cultivars will not ripen fruit in this territory. An example is 'Susquehanna'. In Kentucky, this high-quality cultivar performs well and ripens fruit by late September–October. In northern Ohio, considered the northern part of pawpaw territory, it does not typically ripen before frosts arrive, and the yield is low to zero. Some milder climate (yet northern) areas, such as Maryland, may have slightly different results, but be very careful and do not experiment with your planting.

If you are looking to ship or handle fruit a lot, consider cultivars that have thicker, more durable skins (although this development is still in its infancy). Do you want the fruit to ripen mostly all around the same time? If so, only grow two or three cultivars that generally ripen at about the same time. If you want to stretch the harvest season out to maximize the duration of fresh fruit sales, then plant numerous cultivars that ripen at different times. With currently available cultivars, it is possible to harvest ripe pawpaws from about mid-to-late August to early-to-mid October, within Zones 6 to 8; farther north the season could be about September to early October.

Remember, pawpaws generally grow vigorously under favorable conditions, but cultivars can vary substantially in fruit quality, including size, yields, flavor, texture, and marketability. So, be very careful in choosing your cultivars by dutifully researching multiple reliable resources, talking to pawpaw growers in your state, and if possible, sampling fruits to ascertain what appeals to you or your operation.

A mistake that many inexperienced fruit growers make, and which can prove disastrous, is to become attracted and infatuated by colorful nursery catalog descriptions. The illusory edited images and descriptions often fail to pan out in reality... so be careful and choose wisely both the cultivar and the nursery to make sure they are legitimate.

Popular often common cultivars are usually popular for good reason. These generally recognized high-quality pawpaw cultivars include, but are not limited to: 'Sunflower', 'Overleese', 'NC-1', 'KSU Atwood', 'KSU Benson', 'Maria's Joy', and Neal Peterson's cultivars 'Shenandoah', 'Susquehanna', and 'Potomac'. Beware, however, that many other named cultivars sold through numerous nursery catalogs may produce inferior low-quality fruit that is not very desirable, is not marketable, or is low-yielding. We discuss the known quality, merits, and downfalls of virtually all available cultivars in chapter 11, Pawpaw Cultivars.

Grafted Trees

What is grafting? Grafting is the art of taking one tree, in this case an actively growing 2- to 3-year-old seedling pawpaw, cutting off the top growth, and then carefully surgically attaching a living (but dormant and leafless) piece of twig of another pawpaw tree (called a *scion*), and then wrapping it with special tape so that it seals and heals over, thus fusing the two. This grafted specimen then resumes growing the scion into an entire tree that is an exact genetic replication of the mother tree from which the twig piece was acquired.

Pawpaws graft easily and in good conditions will live and produce fruit after about 3 to 5 years in the ground, and continue to produce fruit for about 15 to 20 years before they begin to decline in production, leaf size, and canopy cover and subsequently die off.[1] Seedlings live indefinitely if allowed to sucker and spread. However, 20 years is a relatively long time and gives an orchard a decent lifespan. Fresh grafted specimens can be planted elsewhere within that time frame to allow for the perpetuation of the operation if desired.

When planting grafted trees, you know what you're going to get. When you plant a 'KSU Atwood' pawpaw tree, you know that in good conditions it will produce copious amounts of medium-sized, sweet tasty fruit with low seed weight and good *Phyllosticta* resistance. However, when planting seedling trees, you don't know exactly what characteristics they will display. If you plant seedlings from parents of excellent genetics, then many of them will most likely be very high-quality fruit producers. Yet for the commercial grower, it just makes sense to mostly plant grafted trees, especially if the goal is to sell *premium-quality fruits* destined to markets or stores. If you were to plant an entire orchard of seedlings, only to find 4 or 5 years later that they produce small inferior fruits, then that could prove to be very problematic, and not easily solved. The best thing to do in that case would be to top-work (graft) most of the inferior trees over to named cultivars as soon as possible. Top-working pawpaw trees have been shown to be effective for that purpose, but will take a few years to regrow large enough to start producing fruit again.

Grafted pawpaw trees generally retail at $30 to $50 each or more, depending on size. Sometimes they can be acquired wholesale for a slight discount. Purchase trees early because, as of 2020, grafted pawpaw trees of high quality are often in short supply and many nurseries sell out fast. Ordering a year in advance is a good idea. Strictly avoid the temptation to purchase only the cheapest trees you can find.

These are usually unhealthy, abused, and may be labeled incorrectly or not at all. Choose reputable nurseries with great reviews and track records, and order when possible 6 months to a year in advance of when you hope to plant. Also, avoid the temptation to go with any random named cultivars out there, thinking it doesn't really matter, or as substitutions for what you really planned to purchase. If you find that your desired named cultivars are all sold out, be patient, check other nurseries, or wait till next year. If there is an excellent local tree you want to reproduce and plant out in an orchard, some nurseries can custom graft and provide trees within 6 to 12 months, or you could learn to graft. Remember you'll need pollination partners (and another cultivar or seedling) of different genetics.

The bottom line is that planting grafted trees is a surefire way to establish an excellent orchard. It is more expensive and you may have to wait a season to get the trees you want, but grafted trees also produce fruit a season or two earlier than seedlings of the same age. Stick to the cultivars you're planning on utilizing and order early. Not all cultivars are available in the nursery trade or not in the quality or quantity you desire.

Seedlings

Why might you want to grow an entire orchard of seedlings? This is an attractive concept to some people. After all, it's the cheapest route, and might seem the most "natural." Some people also believe grafted trees are inferior (they're not!). And, people considering growing pawpaws usually find out before long that they can often get seedlings from wholesale nurseries for as cheap as $1 to $5 per tree. What a bargain! What I have to say about that is this: don't do it if you want to sell premium-quality fresh fruit or want maximum yields and profit. The reality is that wholesale pawpaw trees in this price range are field-grown specimens of random genetic background, are not dug carefully, will have very compromised root systems, and many will fail when planted. Wholesale trees like this are meant for wildlife plantings and reforestation, not fruit marketing. Modern orchardists growing other common fruits to go to market would virtually never consider planting an orchard of seedling trees.

Note: pawpaw seedlings potentially can live much longer than grafted pawpaw trees. When grown in excellent conditions, grafted pawpaws have a productive life of about 15–20 years before they slowly stop producing fruit and go into decline and die. In decent but not optimal conditions, they have about 15 to 20 years of good production before starting to decline. Seedling trees allowed to create suckers and spread can live for decades, possibly a century or more. So, if longevity is a main concern to you, plant high-quality potted seedlings of good genetics, or otherwise integrate them into your orchard operation. It's probably a good idea to give the seedlings their own row or area, so that they can sucker somewhat, if this is what you're after. However, most of us will be plenty content with two decades of solid production from a grove, and will have probably planted another fresh new grove within that time frame anyway to replace the first grove that will soon decline.

Seedlings grown from excellent genetic backgrounds (parent trees) often produce excellent quality fruit. However, unless utilizing grafted trees, it's always somewhat of a gamble because some of those will not prove as good as the parents (though likely not awful like many random seedlings are). Remember, if you want premium fruits, plant premium trees. This is the same path most modern farmers take whatever they raise.

There are good reasons for not growing an orchard of seedlings:

- **Seedlings can have any number of different characteristics that may or may not be conducive to your operation.** However, pawpaws do perform quite well from seed when that seed comes from excellent parents, unlike apples (researchers consider one excellent apple tree from 10,000 seeds started to be about average). But, if you are wanting to market pawpaws as premium fresh fruit, why gamble with genetics and place your bets that it will turn out great? Save years of time and possibly your entire effort by growing an orchard of all or mostly high-quality grafted trees.
- **Seedlings generally take longer to produce fruit than grafted**

trees. Although well-grown seedlings in excellent conditions grow rapidly, grafted trees should bear fruit a year or two sooner than seedlings of the same age. Small seedlings under excellent conditions take at least 4 years to bear fruit, and usually about 5 to 7 years.

- **Most seedlings are derived from wild-grown pawpaws or a random assortment of genetics.** Again, planting these is playing the genetic lottery, gambling that you'll get an orchard full of heavy-yielding, sizable winners, but the chances of that are slim to impossible.
- **Most seedling trees are only available shipped to you bare-root.** Pawpaws can handle this when the nursery does an excellent job and you, the recipient, does an even better one in handling them. However, most wholesale nurseries grow millions of trees a year, harvest them very hastily using large equipment and migratory workers, and they just can't treat each tree very carefully. So, in any batch of wholesale trees, you'll get some that are dead, broken, and toothpick-sized. Container-grown pawpaws perform much better due to less disturbance of their fragile roots. They undergo less transplant shock, recover faster, and produce fruit sooner than most bare-root trees. Both can be successful, but go with container-grown from reputable nurseries.

Always choose the very best planting stock you have access to, or grow your own. This is a well-grown 2-year-old seedling.

So, considering these factors, be realistic and strategic and plant your orchard using excellent grafted trees. Seedlings are fun to experiment with, and after all, every excellent cultivar began as a seedling. Start some seedlings of your best fruits and plant them in the orchard. They'll likely be good pollinators and may turn out to be fantastic fruit producers. That being said, planting a row or two of seedlings from a superior genetic line within your pawpaw grove will help ensure better yields and strong pollination. It's just not the best idea to trust your whole orchard to seedlings, unless that's all you can afford or you're just doing it for fun or experimental purposes. At our nursery, we produce a line of superior pawpaw genetics for just such a purpose, including seeds from 'Sunflower', KSU selections, Neal Peterson's selections, and other excellent acquisitions. So, get your cultivar list together and research reputable nurseries to buy your trees from (or graft your own), and do it early.

6

Maintaining the Orchard

Pawpaw orchards are grown and maintained much the same as most other temperate fruit orchards. Thankfully, pawpaw trees are known to be low-maintenance and resilient to disease and insects. However, they are not invincible and do require some amount of maintenance for optimal health and results. Maintenance consists of:

- Weed and grass control
- Fertilization
- Irrigation, if required
- Pruning and training
- Spraying
- Deer and pest protection

Weed and Grass Control

Before developing a solid mulching system, you will need to control grass and weed growth. Most orchards maintain short grass growth between their trees and within the rows by regular mowing. That works well enough. We use a European BCS walk-behind tractor with a sickle-bar mower attachment, which allows close mowing and close row spacing, due to the walk-behind's small size (not much bigger than a riding lawnmower). This makes the orchard more space efficient than using a much larger tractor-pulled mower. It's also much more difficult and slower. A riding lawnmower or zero turn mower would be a good in-between; it works well if the grass is kept cut and not allowed to get overgrown with brambles or tall weeds, and is much easier to handle and use than a BCS or similar machine.

Some people might enjoy experimenting with animals grazing between rows. We have successfully used moveable electric netting to graze dairy goats in between pawpaw rows. Chris Chmiel of Integration Acres, where goats freely roam his pawpaw groves, discovered that, surprisingly, goats do not forage on pawpaw trees (although they will sometimes rub trees with their horns and this will damage them, which is not acceptable for commercial orchards). Electric fencing could allow cows or other animals to graze safely between rows. In a system like this, you could be simultaneously producing multiple products besides pawpaws—milk, meat, ducks, wool, etc. Or, hay could be cut in the space between rows.

There are many ways to manage the growth in the orchard besides simply mowing down the grass. But unless you carefully design and maintain it, mowing is an easy and efficient way to manage grass and keep everything in check. Designing such systems is beyond the scope of this book, but information can be found in permaculture and sustainable farming publications. Also note that animal manure under the trees may create a contamination risk at harvest time. So it's probably best to remove animals at least a month or more before fruit ripens.

Pawpaw trees, especially when young and establishing, cannot compete with thick sod, tall grass, or weeds of any kind, including weed trees. Pawpaws quickly get overtaken and can be rapidly killed off or stunted by the competition. Woven plastic ground cover or thick organic material mulch and cutting grass down is the best overall way to handle this issue. Even with mulching, though, some weeds will make it into the inner circle of unmulched ground directly next to the pawpaw tree; they should be hand weeded or cut down to ground level with hand snips or a kama tool (serrated Japanese hand sickle, an essential for any organic farmer!). When trees are at least 5 to 6 feet tall, a flame weeder can be utilized with extreme caution. When trees are small, avoid using this as the hot flames can damage any leaves or young twigs. It will also melt plastic tree shelters and ground cover and could ignite dry organic mulch! Chemical herbicides could also be used, but are not recommended and cause cancer in humans.[1]

Mow between your trees at least several times per season, based on the height of the grass and your equipment, not the calendar. This will vary depending on your area and seasonal precipitation. Here in Kentucky where summers are fairly wet (averaging .5 to 1 inch of rain per week), we have to mow at least 5 or 6 times from around early May to October, or about once a month. We let the growth get quite thick and lush before we mow, allowing the seeded legumes to bloom and plenty of organic matter to form, which, when cut, becomes mulch. We are certainly not going for the golf course look, and unless you favor excess work and resource use, you should avoid that as well. When gauging when to mow, remember your equipment's capacity. Our BCS with sickle bar can cut down massive weeds, young shrubs and trees, and grass over 6 feet tall. A zero turn mower shouldn't cut anything taller than 5 to 6 inches and cannot handle anything woody.

Also keep in mind the effect lush growth has on pests. Tall weeds encourage and create breeding grounds for slugs and snails, one of pawpaws' major pests when trees are young, capable of defoliating entire saplings. See chapter 9 for full details on slug and snail control. If they are a major problem, especially in a wet season, consider mowing more frequently.

Mulch

Mulch is a prerequisite, not an option, when attempting to establish and grow healthy pawpaw trees. It can be either natural wood chips or thick straw/organic matter, or can be plastic ground covers. Both work and both have advantages and disadvantages.

As previously explained, pawpaws cannot compete at all with grass sod, weeds, or other thick plant growth, especially when they are small and establishing. The roots simply get choked out, and the trees will struggle or fail. Bare-ground cultivation is an obsolete practice, which is seldom recommended in orchards these days, is almost never done by small-scale growers, is not a sustainable approach, and will almost certainly not give as good results as using thick mulch. Organic mulch materials need not be any certain material but should be free of chemical residues like cancer-causing glyphosate (Roundup) and should probably not be cedar-based. Cedar, as well as black walnut (*Juglans nigra*), is toxic to many species, yet some people claim pawpaw is not affected by black walnut or *juglone*, but if given the choice, why risk it?

Remember, mulches should never touch the trunk of the tree because its decomposition can adversely affect the trunk. It also attracts and fosters sow bugs (rollie-pollies/pillbugs), which are a major pest for pawpaw. If they take up residence on the tree trunks due to mulch contact, they can debark the tree and kill it. See chapter 9 for more information.

A good formula for mulch is 3-3-3: 3 feet wide, 3 inches thick, and 3 inches from the trunk. It can even be applied thicker than 3 inches, but should again, never touch the trunk. If using bark or wood chips, be aware that, when applied fresh, these bind up nitrogen due to bacteria using the nitrogen in the topsoil to break down the carbon in the wood, thus making the nitrogen unavailable to the trees. So, add extra organic nitrogen material on top of the soil (under the mulch) whenever using fresh chips or liquid feed with nitrogen in the form of fish emulsion.

More organic growers are going to want to use *organic mulches*, which give the best results. The main advantages to organic mulches are that they are natural materials from the environment and so they supply abundant organic matter for the topsoil, they trickle nutrients and trace minerals as they break down, and they stimulate soil microlife, beneficial fungi, and earthworms. Big advantages indeed! Also, they are often available free of charge, do not cause any environmental pollution, and are not manufactured. They conserve water and protect the soil from wind and water erosion, and cool it on hot summer days.

The main disadvantages to organic mulch materials are that they are sometimes difficult to obtain in quantity; they are often extremely heavy, difficult to apply and move around the orchard; and they rapidly break down, need annual reapplication; and weeds that readily establish within the mulch layer can be difficult to remove. A fourth disadvantage is that they can sometimes contain traces of pesticides such as glyphosate or fungicide sprays, so be careful where you obtain them. An excellent and usually chemical-free source of wood chips in many parts of the US is any local or municipal tree-trimming operation. Be forward and politely inquire with the local companies if they would be willing to dump chips on your land. When I say be forward, you may have to flag down trucks or visit them on the spot when you see them cutting limbs or trees down. Be brief and to the point. Make sure your dumping site is convenient, very near the orchard, and accessible by an enormous top-heavy truck. If the site is overly wet, sloped, or risky for a huge truck, don't expect to get chips delivered there. Also, be sure to be very communicative about when and where and how many loads you want dumped. Once they find a good and free site to dump chips, they can become overzealous! Beware of chips containing large thorns that can pop tires, such as from honey locust trees and Osage orange.

To remedy one of the main disadvantages of organic mulch materials, that is, their rapid decomposition in wet climates, a thick layer of flattened plain brown cardboard placed under the mulch layer will extend its usefulness and life considerably and will kill most of the grass and weeds underneath. Place the cardboard in a 6-foot diameter (3 feet from the center on all sides of the tree) circular or square shape. Cover with thick 3 to 6 inches of organic mulch. If using straw or hay mulch, use up to 18 inches because it will soon rot and shrink down rapidly to 3 to 6 inches. Just use whatever local organic matter is in abundance; wood chips, newspaper and cardboard found in retail store dumpsters, straw and spoiled hay, muck and animal bedding, grass clippings, rotted manures all work well. Make sure the cardboard is covered in organic matter or it will blow away. Also check with your organic certification rules to make sure cardboard or newspaper is allowed in your production system.

Synthetic mulches are the other option. Always keep pawpaw trees mulched, even if only using synthetics such as black plastic ground cover. Ground cloth is the nursery name for a durable woven plastic cloth that can be cut to size and placed around the trees, held down securely by hammering 6-inch metal sod staples through the cloth into the ground about every 18 to 24 inches of material, 5 or 6 staples per tree.

Ground cloth, although synthetic, is an excellent choice for those that have limited labor, have no tractor with a bucket, lack access to mulch materials, are older, physically limited, or simply want to do things the easier way (without resorting to toxic herbicides), or simply have a lot going on besides their orchard. It does not give as optimal results as the soil life-stimulating and nutrient-providing organic mulch. However, it completely kills and suppresses grass and weeds and is cheap, easy to apply, and long-lasting. There are no harmful chemical residues like with herbicides. Good-quality woven material can last for 10 years; you only need to apply it once or twice in the life of the orchard. Don't confuse this ground cloth with the cheap, garbage bag-like plastic the big-box stores sell to use in flowerbeds or with other soil-smothering and hard-to-use felt-like material. The real cloth has to be purchased from a nursery supply company or farm supply warehouse. Check the resources section for suppliers.

In our orchard, we utilize ground cloth and, on a smaller planting, organic wood chip mulch and cardboard. Both give good results. The way our farm is set up and the fact that we do not own a tractor prevents us from being able to mulch our 400 some fruit trees by hand with organic mulch. It's just not going to happen. So, we have found the ground cloth to work great. Remember, you're only covering a small 3- to 4-foot diameter area around each tree. This is not killing the soil or the soil life! It's only preventing vegetation from growing in that little spot around each tree, and this makes all the difference in the ability of the tree to establish and thrive early on. You have to use something, whether organic or plastic, and you'll have to judge what is right in your situation.

When using black plastic mulch, you can either cut individual squares for each tree or just lay down a long line of it. We prefer individual squares, but for the sake of completeness, here is how you can use long lines, which gives the greatest amount of weed and grass control. In completely non-windy weather, start about 4 to 5 feet from the first tree in the row, weigh the plastic down with blocks or hunks of wood, and then unroll it about 4 to 5 feet past the last tree. Before unrolling, position it next to the row of trees (so you can unroll it). After it's unrolled, cut a slit into the plastic that will allow you to slide it around each tree, like a collar. Position it, and hammer the staples in place along the edges, using 5 or 6 sod staples per tree. To avoid smothering the tree, cut out a 6- to 9-inch circle around each one, large enough to water or apply organic fertilizer, but small enough that the weeds are covered and don't proliferate in the opening. It's difficult to get the plastic perfectly lined up, so just do your best and never cut the slit before placing it directly next to each tree, one at a time. Use black plastic that is 4 to 5 feet wide, available in 100- or 200-foot rolls. Although not the most aesthetically pleasing, it works great, is very fast and easy, and your trees will grow and succeed with it, which is the main objective.

Fertilizer materials should be placed within this hole in the plastic (not on top of it) so roots have better access. Drip irrigation is generally placed under the plastic mulch. And, be aware that if you cover the plastic material with organic materials, either to look better or to try to stimulate the soil life, these materials will decompose on top of the plastic and weeds will thus germinate on top of the plastic and will root through it, thus eliminating most of its benefit and causing a mess. So, it's better to keep it somewhat clean, unless you want to have a small ring of mulch around the tree within the hole you cut. Just don't let any material touch the trunk, like always. Overall, using single squares of the material around each tree works great, looks better, and is probably the better option for larger operations.

Once pawpaw trees are established (at least 5 to 6 years old), use a low-growing, aggressive, managed understory layer of such plants as clovers, comfrey, lespedeza, wild violet, or other non-grass ground covers in place of any mulch. Herbal permaculture-type understory plant guilds are fine too. Even when trees are mature, having thick grass growing right up next to the trunk can't be optimal.

Weeds

As we learned, thick mulch takes care of most weed issues in the orchard. Before planting, simply mow or scythe down all plants growing near the planting hole for easier digging. We usually remove any sod by hand with a pickaxe, scraping it away shallowly. After planting and deeply watering the new tree, install weed control, as discussed (cardboard and mulching thickly or use ground cloth material), and that's about it. There are a few troublesome weeds worth mentioning that are found in the eastern US and beyond.

Johnson grass (*Sorghum halepense*) creates a thick, ropey rhizomatous mass underground and can be very aggressive. Its takeover of a field is usually a strong deterrent to orchard establishment. However, thick and wide mulching and mowing should keep it under control. Michael Phillips says if a field is covered in Johnson grass, it may be better to choose another site. I have not found it to be a very big deal; mowing it keeps it under control. When it wilts from frost, it becomes toxic to livestock; otherwise they enjoy grazing it, and it makes good hay.

Bindweed (*Convolvus arvensis*), resembling and related to morning glory, is an aggressive, long-lived, and persistent annually appearing perennial vine that can quickly climb up young trees or tree shelters and smother them. To control, cut it at the base to kill the tops, preferably before it flowers. If you have bindweed, check on young trees at least every 2 to 3 weeks and remove it as necessary. It also loves to climb chicken wire cages. Simply yank the cages upward on their stakes to dislodge and rip the vines and then replace the cage. String trimmers work well too. Check inside the cages to make sure no bindweed is climbing the trees themselves and smothering them. Try to keep the trees and cages clear of bindweed so it does not go to seed.

The same advice above goes for morning glory. Bindweed seeds can live over 30 years in the soil, and has perennial roots, so control is important. Morning glory is also troublesome but fortunately does not make perennial roots. It also must be kept off of the trees.

Wild mulberry (*Morus alba*), while also a wonderful tree and fruit producer, can become a serious weed when not managed. Birds feed on local mulberries and then often land on tree shelters or trees and defecate, thus depositing the seeds. These sprout next to the pawpaw tree and within a couple of short years can smother a young tree if the mulberry is not removed when it is still very small. When 4 to 5 feet tall, they can be very difficult to remove; even cutting them down to the ground often just stimulates regrowth, and digging may not be possible right next to pawpaw trees without damaging them. So, clearing around them, weeding, and mulching annually will help, as well as removing nearby wild mulberry trees if possible.

Wild Rose (*Rosa multiflora*), **Japanese honeysuckle** (*Lonicera japonica*), and **wild blackberry** (*Rubus* sp.) can all be problematic and should be handled the same way as mulberry. Pull them up when young and allow to dry in the sun.

Zoysia grasses are rhizome creepers and can creep right onto your pawpaws and invade their root zone. It colonizes organic mulch quickly unless controlled.

Suckers, while technically not weeds but part of the pawpaw tree itself, emerge from underground and can get out of control in a neglected or not very well-managed planting. They are very easy to control through mowing, scything, sickles, string trimmers, and the like. Remember, on grafted trees any suckers that come up are not the named grafted cultivar and would represent the seedling rootstock genetics. If you only had one pawpaw tree and it was grafted, its suckers could be left to mature and pollinate the grafted tree. However, planting other

Pawpaw is strongly suckering, sending up many little shoots that, if left unchecked, can overrun the planting and overpower the grafted portion of the tree. These are root suckers and are not of the same genetics as the grafted tree portion and should be removed via scythe, mower, string trimmer, etc.

trees would be a much more viable solution to lack of pollination. Left unchecked, suckers will create a pawpaw thicket—not what you want for premium fruit production. Cut them down to ground level with hand snips when they emerge right next to the tree. Otherwise mowing keeps them down.

Fertilization

Fertilizing pawpaw trees is a new agricultural "science", and there are a lot of unknowns. Fortunately, pawpaws are pretty flexible as long as a few things are understood. They are heavy feeders, requiring yearly applications of a balanced fertilizer in order to perform, as well as a heavy source of nitrogen. Neal Peterson says pawpaws like abundant potash applications. Kentucky State University (KSU) recommends 1 ounce of nitrogen per tree per year, 2 ounces after 2 years of growth.[2] March, May, and June would be good times to fertigate and/or fertilize your trees.

Many of you will want organic recommendations, which is how we farm as a Certified Organic operation. We believe pawpaw trees are similar to blueberries (and in reality, almost everything) in that they perform best with organic growing conditions and organic sources of nutrition. Why wouldn't they? Every plant does best with organic fertility inputs. Just ask every thriving forest ecosystem. KSU recommends NatureSafe organic fertilizer, which is a turkey manure/feather product with an analysis of 10-2-8. Other granulated organic fertilizer blends or granulated/pelleted chicken manure work great as well and are usually our main fertilizer application.

KSU recommends broadcasting fertilizer under the pawpaw trees before bud break in early spring (March–early April usually), applying at a rate of 1 ounce nitrogen (N) per tree the first year after planting, then 4 ounces N per tree (about 50 pounds per acre) from years 2 to 5. Finally, in years 6 and beyond, a rate of 5 to 6 ounces N per tree is recommended.[3] This number is describing using "actual nitrogen," so you will have to translate the NPK number on the bag when using fertilizer products. If confused, contact your local agriculture extension office about translating actual nitrogen numbers to your particular fertilizer's NPK numbers. It will translate to several pounds of organic fertilizer product per tree. Online there is a handy calculator for this purpose created by the University of Georgia.[4]

We use a simple approach that works well with organics because organic fertilizer is much less likely to overload plants, especially trees. Caution is still advised not to overdo it. Our particular site is fairly fertile, high-quality silt-loam soil. We give each tree approximately ½ pound of granulated chicken manure the 1st year in the ground a month or so after planting, and then 1 pound the 2nd year in the ground in mid-April. Later we give another 1-pound application in May or June. We also apply a diluted fish emulsion at least 2 times per spring/early summer, around the same time as the chicken fertilizer (about 1 cup of fish emulsion to 4 gallons of water, each tree getting about 1 gallon of this solution). We also spray the orchard about once per month (April–August) with a kelp extract and water solution, the minimum being 1 or 2 times per growing season. Kelp extract provides a wide array of trace minerals, including boron and potassium, and some beneficial plant hormones. This regimen gives strong results and fast growth. You should increase the chicken fertilizer by 1 pound per year the tree has been in the ground. For instance: year 2 = 1 pound, year 3 = 2 pounds. Up to 4 to 5 pounds per tree should be sufficient at maturity. Very infertile soils may need more, and should receive heavy mulch and compost.

I would like to explain why organic fertilizer is very superior to chemical salts that stimulate plant growth (aka chemical fertilizer). We all know salt, being antimicrobial, is bad for soil life. Too much makes it very difficult for plants and soil microbes to live. We see this in the salination of heavily fertilized and irrigated farmland in California and elsewhere. Organic fertilizers, such as manures, are not just NPK numbers and salt; they are a matrix of microbe-stimulating organic matter within which is powerful nitrogen, potassium, phosphorous, and also trace minerals. Manures, agricultural residues, and compost contain a lot more than NPK. They stimulate soil life and earthworms while gently and gradually yet effectively stimulating growth. They do not wash out of the soil in heavy rains like synthetics can (leading to toxic algae growth in waterways and ponds). It's hard to overdo organic fertilizers, and if you do, it usually just leads to excess growth and does not kill the tree or plant. It leaves the soil healthier, not depleted and full of synthetic residues like chemical fertilizers do. They cost more than chemical fertilizers and take more material to get the job done, but many easy options are available these days. Isn't our environment and soil worth it?

The best way to ascertain if your fertility program is working is by monitoring tree health and vigor. Leaves should be large, bigger than your hands, and vibrantly deep green. Veiny or yellow leaves usually mean a nutrient deficiency. Shoots should be long and vigorous, displaying at least 16 to 24 inches per year of growth the first years of establishment. Stubby, short shoots usually mean a weak tree or lack of fertility. Flowering should be abundant. Signs of lack of fertility can be quickly corrected in an organic system, and over-fertilizing is not deadly with organics as it can be when using chemical fertilizer.

We find pawpaw trees respond to immediate fertilization, from seedling to juvenile trees and then afterward. Poor fertility leads to trees being stunted, showing such deficiencies as yellow veiny leaves, or they simply grow extremely slowly. In addition to the granulated fertilizer, pawpaws also respond very favorably to well-rotted compost or cow manure thickly applied on top of the soil, under the mulch or plastic mulch layer. This seems to be very stimulating to them.

Liquid organic *fertigation* using kelp and fish emulsion works extremely well and can be applied 1 or 2 times per month from March to June. This can be put through your irrigation system or diluted in a 55-gallon barrel and fed to the trees manually via buckets or a hose. Make sure all local laws are followed when doing fertigation if utilizing municipal water supplies. Liquid fertilizers can also be applied as a foliar feed via a fertigation attachment on a garden hose or in a backpack or vehicle-attached sprayer system. Kelp extract can be added to the fish emulsion for beneficial micronutrients and hormones, as well as any liquid organic pesticides/fungicides such as "Bt" products and strained compost tea. It is recommended to not apply any nitrogen to the trees past late June to avoid stimulating excessive late-season growth that will not harden off properly before freezes occur. So, make sure to fertilize very heavily from March to June to get strong, healthy growth. KSU defines this as 16 to 24 inches of shoot extension per year in establishing trees and about 6 inches in established trees.[5] Keep these numbers in mind when evaluating your grove.

One final note on fertilizing is that pawpaws are not extremely picky. If you simply threw plenty of local rotted manure and piled mulch around the trees, kept them moist, and cut the weeds, I feel confident that pawpaws would still perform very well. Their (still) wild nature makes them quite adaptable and resilient. Only get as technical as you feel necessary and base the inputs you use more on the trees' response via production, measured shoot growth, appearance of the leaves, etc. and less on numbers on a soil test or what someone else says.

Irrigation

In most orchards, fruit trees grow faster and stronger and are more productive when irrigation is provided. And, in many areas like the Southwestern states, fruit trees will often only survive if irrigation is provided. However, irrigation of pawpaw orchards is a practice that is not very standardized as of yet. Primarily, pawpaws are grown in the eastern US which generally receives plenty of year-round rainfall, and so irrigation is not absolutely required here on most sites. However, it's wise to implement or at least have available some form or means of irrigation. You do not want to place all your bets on their being plenty of rainfall the year you establish your orchard and then witness it ruined by a drought. Unfortunately, this is what happened to famous pawpaw breeder Neal Peterson, and he lost his just-planted pawpaw orchard (and his farm) because of it.[6] He had planted an entire orchard of saplings on the top of a ridge in West Virginia, and that summer happened to be an extremely hot and dry one, and all his non-irrigated pawpaws died, and he lost the land. Don't let this happen to you! On our farm, 2019 proved to be a brutally hot and dry summer after June and July's heavy rains. We had to water over a hundred newly planted pawpaw trees several times after a month of virtually no rain and temperatures in the 90s. Make sure you can irrigate your trees!

Irrigation is important to obtain optimal yields of fruit most years, and also is crucial when establishing a new tree planting. Don't assume there will be plenty of rainfall when establishing a planting. Have plans in place for getting the trees supplementary water if and when needed. Irrigation is absolutely required when growing pawpaws in higher elevation and drier locations. Setting up irrigation can be as simple as placing ¾ inch or ½ inch *orchard tubing* along the tree rows and setting up two *drip irrigation emitters* at every tree. Orchard tubing and emitters are commonly available from most farm supply warehouses. See Resources section for sources. You will need a little tool to cleanly puncture the orchard tubing (on one side only!). We actually use the metal wire of a contractor's or surveyor's marking flag to do this. Then the emitter can be installed into the hole, which is not always easy. A pair of channel locks comes in handy to hold and help push the emitter into the hole, but be careful not to break it in the process, as they are made of plastic. Incorrectly punctured holes will not drip, and incorrectly placed emitters will pop off and cause heavy water leakage. The emitters should make a "pop" sound when inserted correctly into the tubing and be able to be rotated without falling off.

Connect the orchard tubing to a hydrant or water source. Strong flows may need a pressure reducer attachment, and a filter is always a good idea. On our farm, we use a filter but don't need a pressure reducer. You'll know you need a pressure reducer if your irrigation tubing blasts off the hydrant or busts apart. You could also try not turning the water on 100%, starting it at 50%. Irrigation lines have trouble delivering proper water pressure and flow uphill for very long distances, so try to arrange water to flow down grade, and not up, if possible. You may need to install several hydrants for large orchards. Hydrant lines split off your main water line via an underground "T" and cost around $120 to $200 plus pipes and installation costs. Make sure to get a frost-free, or nonfreezing hydrant, and completely drain it every autumn before freezing weather by detaching all hoses and attachments from the hydrant unit. Locking it is not a bad idea either; the little holes on the handle are for sliding a padlock onto it.

It's best to automate the irrigation system for consistent water delivery, ease of operation, and so you never forget to turn it on (or off!). You can easily set up the irrigation to automatically turn on and off via a digital battery-powered timer that attaches to the hydrant and the orchard tubing, or you can install a permanent solenoid system. I recommend setting up a permanent solenoid system that is tied to the electrical grid. A professional may have to install this, unless you are proficient at electrical wiring and setting up irrigation systems.

We've used various brands of the digital battery-powered timers, and they are convenient, easy and fast to install, and usually work well. However, they can also stop working suddenly without warning. You may find that your trees are suffering from lack of water, and your battery-powered timer has stopped working, perhaps days or weeks ago, without warning. These devices are cheaply manufactured and are sometimes unreliable. If you do choose to use them, check them once or twice per week to make sure they are still functioning. Use fresh name-brand batteries every season. Purchase only the highest-quality professional timers and install them so they are not touching the ground. They have two hose-type attachments, one male and one female, for inflow and outflow. We have installed them directly on the hydrant and attached the orchard tubing onto the other (outflow) end. If using a pressure reducer, put the reducer on the hydrant, and attach the timer on the other end of the pressure reducer. Attach everything together very tightly with channel locks so there is no leakage, and you may need to use Teflon tape or plumber's putty/paste. Having some shelter over the device is a good idea to reduce sun exposure that degrades the plastic casing and the LCD display. Strap it to the hydrant with metal pipe strapping so it does not move, or it will certainly break off. Make sure the irrigation valves on the orchard tubing (if any) are open, and keep the hydrant on all season (with the handle in the up position). The timer operates a small valve inside the unit that opens to allow water to flow and closes the valve to stop the flow, so there's no need to manipulate the hydrant or main valves. At the end of the season, around late September, turn off the water and remove the timer from the hydrant or main line before freezing weather hits. Remove the batteries, clean and store the timer indoors out of freezing weather. Don't expect more than one or two seasons from a timer. They generally cost between $30 and $40 and can be purchased online or at home-improvement stores and some hardware stores. Don't buy the cheapest one you can find! Remember, the timer is like the lifeline of the orchard in some situations. Setting up a permanent solenoid system is a much better option. That is beyond the scope of this book to explain, but that information is easily obtained.[7] It takes more work and resources to establish, but will serve your operation much better. A good solenoid system will cost around $150 to $200 plus labor charges to have it wired, plumbed to the main line, etc.

Some people get by hauling irrigation water in containers on 4-wheelers. This is labor-intensive, but may only be necessary the first season or two when establishing your trees, if you live in a heavy-rainfall area (40+ inches a year). Smaller plantings up to an acre or two could be watered by hand with very long connected garden water hoses and a spray nozzle (though this would get old pretty fast!). Mist or sprinkler systems also work fine, but they use excessive water that ends up watering grass and weeds as well. Irrigation water derived from the municipal water supply can get expensive, so be careful with what you choose to do.

Drip irrigation is hands-down the most efficient way to go. Watering during very hot and very dry spells can help prevent leaf drop and subsequent sunburning/splitting of fruit and loss of crop. Too much irrigation may lessen fruit quality, break branches due to heavy fruit loads, and be unsustainable financially. Consider rainwater catchment off a barn, large shed, or house as a sustainable (and free) water supply.

KSU recommends setting up two emitters per established tree set to deliver 1 gallon of water (in one hour in the morning or at night) per day, for a total of about 240 gallons per year, per tree (on well-draining soil).[8] That's delivering each tree about 1 gallon per day, or about 30 gallons per month, for 8 months a year. Kentucky State University has begun hanging their irrigation lines about 3 feet off the ground, attached to metal "T" fence posts which are set in the ground, one placed at each tree in the row. This works well for them, after finding their irrigation emitters got clogged with dirt and debris when left on the ground next to the trees. The water simply drips down and lands next to the trees, soaking into the soil.

If minimizing irrigation is a goal and you are planting in a relatively high rainfall region, such as Kentucky, then the irrigation could be set up to deliver 2 gallons per tree per day to be delivered only during the hottest four months (May through August), or about 120 days of irrigation per year. If March or April during flowering time were very dry (a very rare occurrence in KY), then irrigating at these times might be important for optimal fruit set. If May or June were very rainy, then irrigation may not be needed and could be turned off in order to save money. Check soil moisture levels with a tensiometer or other moisture meter if necessary. Too much water may cause problems such as diluting fruit sugars and oversizing the fruit or flooding the trees to death. Too little moisture may lower yields or completely kill off a new planting, so be careful. This also calls for some common sense. If the orchard seems bone-dry, it probably is. If the orchard is flooded or there's standing water, then stop irrigating! Use a moisture meter if in doubt. Generally, if the top few inches of soil appear very dry and dusty, irrigation may be beneficial. Thick mulch and high organic matter in the soil reduce watering needs and keep soil moisture high and even. Organic matter acts like a sponge in the soil; it holds moisture when it comes, banks it, and then slowly releases it when dry.

Once again, irrigating pawpaws is currently not a highly developed science, but because they make deep roots and are native to the eastern US, they are resilient and adaptable when grown in similar conditions. To summarize, in a moist region (35 to 40+ inches of rain per year), on a good site, irrigation after the first season or two is optional, and likely beneficial, but may not be necessary if summer rains have been adequate. Thick mulch and plenty of organic material in the soil facilitate balanced soil moisture levels. The first season of establishing pawpaws in an orchard setting is a fragile time for the trees, and they should be carefully monitored for soil moisture unless it's obviously a very rainy summer. Irrigation can also increase yields when the trees are developing fruit. Don't overdo irrigation. Pawpaws are hardy trees and do fine with no supplemental irrigation in their native habitat, which includes shade, high levels of soil organic matter, and ample rainfall. However, do not trust that rains will be adequate, especially when establishing a new planting, and have irrigation or water supplies ready to roll if needed.

Irrigation Troubleshooting

No water is coming out of the hydrant. Check to make sure the water is on at the main. Contact your water company if you need help locating the main line. If it is, the waterline could have been broken underground by a heavy vehicle or ice event, especially if it was not installed to code (buried below the local frost line). The hydrant could be clogged or broken. Check the hydrant first to make sure it is in sound condition. Make sure any electric timers or solenoids are functioning and not turned off. Turn off all household faucets and other water outlets connected to the main; if the meter is turning, you have a leak or broken pipe. Get this fixed immediately. If the hydrant is leaking, replace the hydrant and see if this solves the problem. When replacing a hydrant, wait to backfill around the hydrant until you know that was the issue and it is resolved.

My emitters keep popping off or the line is bursting, or the line busts off/comes loose on the main or hydrant. Make sure the emitters are installed properly. If the lines were not punctured properly or fully, the emitters will leak or pop off. If that is not the issue, then you need a pressure reducer. You could also partially close the hydrant to 50% or 75% flow and see if this works for you.

My emitters are clogged and not dripping. You need to flush the entire line by uncapping the ends of the orchard tubing lines and turning the water on for 1 to 2 minutes. Then recap and install a filter at the hydrant. If individual emitters are still clogged and not functioning, and they cannot be cleaned, then pull them out, install a "goof plug" repair piece (large side of the plug, not the smaller side) pressed firmly into the hole in the tubing until you hear it pop. Then make a new hole a few inches away and install a fresh emitter. Clogged or broken emitters cannot be pulled out and replaced with a new emitter in the same hole. Removing emitters tears the hole, and that spot will always leak heavily if a new emitter is installed there. Cap off previously used hole with a goof plug (large end) and a new hole made with the emitter installed there only.

My pressure is very weak and the far end of the line is not dripping enough (or any) water. You may be running the lines uphill. Install another hydrant and run the lines downhill. Make multiple lines and have them irrigate at different times via multiple timers, and not one extremely long line. Make sure also there is no underground leak from a broken pipe or hydrant. Keep the hydrant in good condition and unhook everything from it before freezes, including hoses and attachments. You may need to install an electric water pump or find another source of irrigation water.

Pruning and Training

Pruning is also a new science with pawpaws, and there's a lot we currently don't know. When planting, it may be best to thin the trees to the terminal (central upright) shoot and remove any very low side growth below 2 feet or so. This may direct more of the tree's energy toward height and developing a strong thick trunk. This is standard practice with apples and pears (those are pruned clean to the trunk below 3 feet). Also, some sources recommend spreading branches to make physically strong crotch angles (60° to 90° from the trunk).[9]

Good crotch angles generally stimulate heavy and early flowering in species such as apple and pear, and this may prove true for pawpaw as well. We do know that pawpaw trees generally do fine with zero pruning. However, often the branches will form poor crotch angles (very acute or parallel to the trunk, similar in appearance to a U or a tuning-fork shape). This can and often does lead to eventual major limb breakage in unmanaged trees.

Limb breakage occurs mostly during high winds, ice storms, or very heavy fruit set. So, training young trees to produce branches in the classic 60° to 90° crotch angle from the trunk is highly recommended. This can be done carefully when the branches are very thin and supple, utilizing clothespins or tree spreader devices. Clamp the clothespins around the young trunk so that they hold down the shoots at a roughly 60° to 90° angle. This only works the first season when the shoots are flexible enough. After that, more substantial spreaders will be needed. Pawpaw wood is very fragile, so be slow and careful with this.

Removal of the smaller shoot should happen as soon as it is noticed, if it is too rigid to train to a better angle. Trying to train a rigid shoot that is too firmly established will cause it to split the crotch of the tree. If this happens, remove the smaller shoot and tightly tie up the split portion with electrical tape and/or twine. It will heal back together just fine if tied back together immediately, before the tissue dries out. Remove the wrapping after 6 to 8 weeks.

Adolescent multi-trunk pawpaw (about 5 years in the ground) showing strong bilateral branching pattern that resembles the classic menorah shape. If the tree does not have this shape, it is likely not very strong or healthy, or may have been heavily pruned. Note glowingly healthy green leaves with wide margins, understory ground cover of ground ivy, and zero grass competition.

Pawpaws do not handle pruning of large limbs very well (anything larger than 3 to 4 inches). They are slow to heal and do not heal over pruning cuts easily. So, it's important to prune what needs to be removed as soon as it is noticed and when as small as possible. This includes thicket-forming suckers as well as shoots originating below the scions on grafted trees (rootstock suckers).

KSU recommends topping pawpaw trees at about 15 feet to encourage branching and easier access to harvesting fruit, and this seems a very good practice. Another professional grower says he keeps his trees at about 7 feet so there is no ladder work at all. Anywhere between 7 to 15 feet is reasonable. We prune to about 7 to 8 feet. This keeps maintenance and hand harvesting easier, and the trees become shrub-like.

Any trees or branches diseased or damaged by Asian ambrosia beetles (see chapter 9) should be removed and burned. Any bad crotch angles (these have a tuning fork or U shape) should be thinned to leave only the biggest branch or leader, or if young and supple, the smaller one could be forced downward to a 60° to 90° angle from the trunk, thus becoming a noncompeting scaffold branch.

Certainly keep any broken, crossing, or rubbing branches pruned, and remove suckers around the bases of grafted trees and seedlings that are being kept to one trunk. With grafted trees, any shoots below the graft union must be removed as soon as possible as they are rootstock shoots and will overcome the grafted portion rapidly. On young grafted trees, these can soon outpace the growth of the scion and suppress the establishment of the grafted growth. This is highly undesirable and can ruin the grafted tree. Check all your pawpaw trees at least 2 or 3 times per season as they are establishing and make sure no growth below the scion/graft union is sprouting. Remove with a sharp pair of hand snips close to the trunk. Once this happened on a small grove we maintain in Louisville, a few hours away from our farm. It

Dormant Season Pawpaw Pruning

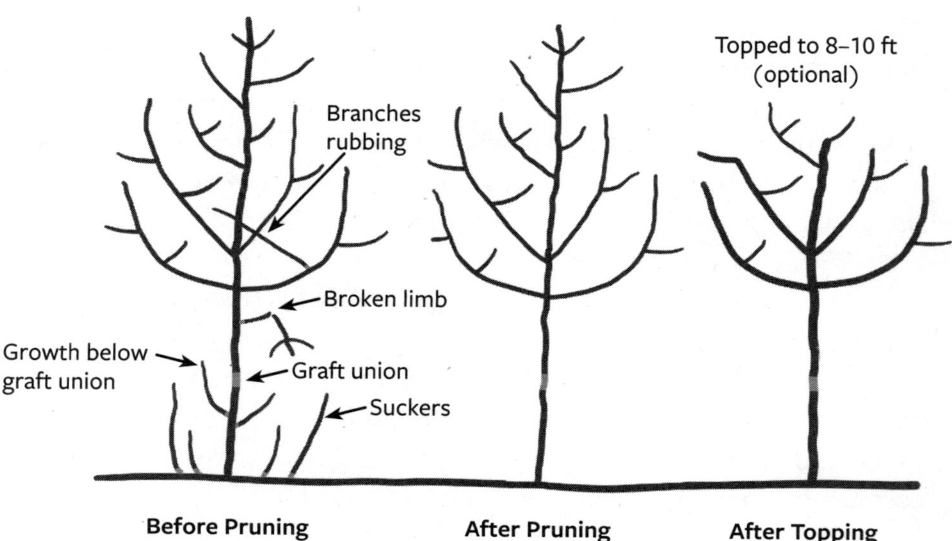

got out of control in my absence and kept growing upward into a second trunk, such that I did not know for sure which was the grafted (desirable) cultivar growth and which was the rootstock sucker; they appeared exactly the same except for a slight difference in leaf characteristics. Eventually the smaller of the two flowered profusely (showing *precocity*), and the other did not flower at all. I hand-pollinated the blooms, and the jumbo-sized, unusually high-quality fruits per the cultivar description clued me in on which was which; I subsequently removed the larger branch. The tree recovered from this, but it slowed down its establishment and left a large pruning scar on the trunk. This caused long-term harm, so be diligent about checking young trees for any amount of growth below the scion or from the roots, and remove immediately when it is as small as possible.

When pruning, always be sure to keep your tools extremely sharp and clean. Make any cuts clean and close to the trunk or close to the remaining branch, leaving *no stub*, but retaining the *shoulder* (the lumpy ring of connective tissue between the trunk and branch) if it is present. Use *bypass pruners* only and never anvil pruners (anvil pruners are only for dried, dead growth). See the Resources section for farm supply sources.

More research, trialing, and time will tell if training and pruning pawpaws more akin to apple trees will be beneficial. Theoretically, training limbs to 60° to 90° angles and creating a scaffold system of fruiting branches akin to apple and pear could prove very beneficial and increase fruit set and reduce the time it takes for trees to initiate fruiting, as it does in many other fruiting species. If you experiment with this and get conclusive results, let me know!

Spraying

Most fruit tree grow guides will have long lists of sprays for everything from mildew and rots to insects of every description. However, as of 2021, no chemical sprays are officially registered for *Asimina triloba*. Thankfully, pawpaw's natural resilience and chemical defenses make this mostly unnecessary. However, some sprays are registered for very similar species, so check with your local laws and regulations if you want to spray. Chapter 9 has in-depth information on *organic sprays* that are potentially effective for various issues.

For general health and vigor of trees, *health sprays* throughout the growing season, consisting of fish emulsion, kelp extract, and compost tea, are all very beneficial in providing trace minerals and helping prevent deficiencies and, possibly, *Phyllosticta* fungal infections. See chapter 9 for more info on diseases of pawpaws and organic solutions.

7

Harvesting Pawpaw Fruit

Pawpaws destined for *premium fruit sales* must be harvested carefully by hand. This is one reason keeping trees pruned to 8 to 12 feet tall is highly recommended. Unless you are young and fit and don't mind lots of sketchy ladder work, keep them conveniently pruned.

Thinning leads to the highest-quality fruit. Removing some of the small immature fruit early in the season allows individual fruits to attain larger size and leads to only one fruit developing per peduncle, or stem. These can be harvested with the peduncle stem still attached; they do not need to be torn from a multi-fruit peduncle, a process that tears off a small piece of the rind, exposing the inner flesh, which bacteria and fungus can easily and swiftly enter and infect. Fruit that is harvested when firm-ripe sometimes heals and seals over this harvest wound, but soft-ripe fruit does not. Any fruit that has a small wound from detaching from the peduncle will only keep at room temperature about 2 to 3 days before rotting begins. Discoloration and/or visible mold at the wound site are indicative of rotting and such fruits should be discarded.

Thinning the fruits, however, is not absolutely necessary and is labor-intensive, taking about 20 to 45 minutes per mature tree. It must occur within a few weeks after flowering, when fruits are still tiny. There are no chemicals or other ways to thin pawpaw fruit, unlike with apples. A *June drop* occurs with pawpaw, wherein about 20% to 40% of the young fruit clusters are naturally shed off the tree, dropping to the ground. This happens also with peaches and apples. It's a way the tree manages its own resources. So, don't count on every single cluster bearing fruit. Kentucky State University recommends thinning before June drop occurs, when fruits are about half an inch long.

Harvesting Pawpaws

By the end of June, fruits should be about golf ball size. Nearly all of them should reach maturity. The first season pawpaw trees begin to fruit, around year 3 to 5, expect approximately 5 to 10 pounds per tree. The next season, this can double, eventually reaching around 30 to 40 pounds in a good season. A good tree yields close to a bushel, about 35 pounds, of fruit per season. KSU has observed that the weight of the yield will be about the same whether the tree primarily yields jumbo-sized fruits or dozens of small fruits.

Pawpaw fruit grows rapidly but takes between 4 to 7 months to mature, depending on cultivar. Early-season cultivars may begin ripening fruit by the first week of August, or slightly before, in the main pawpaw growing range. Late-season cultivars ripen in late September or early October. The main pawpaw season is always September. However, some years, depending on weather, typical harvest times can vary 1 to 2 weeks. In 2017 in Kentucky, warm spring weather came on early, pawpaws were ripe in early August, and all cultivars were done by late September. Of course, in the northern range of growing territory, these dates will vary and ripening will generally be 2 to 4 weeks later for all cultivars.

Pawpaw fruits have a curious seam-like line running lengthwise from the stem end to the tip of the fruit. When very ripe, the fruit easily falls off in the wind or with a slight shake of the tree, usually tearing from the main peduncle. It is considered proper harvest technique to only ever clip the peduncle free from the tree and very carefully handle the fruit; never attempt to shake branches to dislodge fruits, which will certainly damage or bruise them and/or make them unmarketable. Shaking the tree also dislodges underripe fruit that will not ripen, ever. Some growers only market fruits that they hand-harvest and discard any that touch the ground, due to food-safety concerns.

Those solely looking to process pawpaws into pulp could put down tarps or straw and gently shake fruits free of the peduncle, placing any hard ones aside to further ripen (or discard if they do not).

Quality pawpaws reach large size, 12 to 16+ ounces, and jumbo fruits can sometimes weigh 18 to 24 ounces or more. Pawpaws that produce only small fruit, 2 to 6 ounces, are not worth growing. Ripe pawpaw fruit gives to a soft press, like a ripe avocado or peach. Don't press too hard or you'll bruise the fruit and tear the skin. Ripe fruits are also noticeably fragrant, and some cultivars will have a *color break* or coloration to a lighter green to slightly yellow, though most do not display this tendency. When harvesting for premium sales, feel each fruit for this softness. Make sure all your harvesters are doing the same, or you'll end up with bushels of hard, unmarketable, wasted fruits that will never ripen. It's not a super-fast process to get premium-quality fruits, and that's part of the reason you'll be asking premium prices for them. In 2019, this is generally $5 to $10 per pound. To compare, high-quality apples in 2019 were about $2 to $3 per pound.

When harvesting, you will need a pair of sharp hand snips (pruners), gloves, and cushioned harvest boxes. A vehicle to carry the boxes from one tree to another is also very helpful; a four-wheeler or a four-wheel drive farm truck also works. Boxes should preferably be those designed for transporting fruits such as tomatoes or peaches. Cardboard ones work, and can be obtained for free at grocery stores. More appropriate are the modern black plastic produce boxes. These are sturdy, reusable, stackable, and collapsible for easy storage. These can be purchased new or found on Craigslist or at flea markets wherever produce is sold, and are usually about $2 to $4 each. Only buy used boxes that lock properly and do not collapse or fall apart on you. Line them with clean soft towels, burlap, or foam. Place each fruit carefully in only one layer per box to ensure the best quality and the least amount of bruised or broken (unmarketable) fruit. They are tender and delicate. Any destined to go out of state should be shipped that day, or put into cold storage and shipped within a day or two only. Ripe pawpaws, picked when firm but adequately ripe, will keep in cold storage for 3 to 4 weeks in good shape. They will then ripen fully when brought into room temperature environments. When refrigerated in this manner, fruit quality can be good but will never be as optimal as when fruits are tree-ripened to maturity. But, alas, in this day of shipping and long-term storage of produce, optimal quality is not always a possibility. Soft well-ripened fruits can only be kept in cold storage about 1 week before they start to spoil. At room temperature, firm-ripe fruits ripen in about 1 to 3 days and last about 4 to 7 days before rotting sets in.

Again, if you are harvesting pawpaws for pulp production, lay down clean tarps, very gently shake the trees, and pick up fallen fruits; because they will be processed soon after, bruising and small abrasions don't matter as much. Make sure this is compatible with current food safety laws in your area. When shake-harvesting the trees, always be very gentle, and if no fruit falls, then don't shake harder! It's not ready yet! Think of it as gentle bumping of the tree, moving it like a breeze would do. Shaking hard will dislodge immature fruits and ruin the harvest. Store the harvested fruit in cold storage if they are not going to be processed for a day or two.

8

Tree Propagation

There are numerous ways to propagate pawpaw trees. Different methods are called for depending on why, how, and when you are attempting this. In short, pawpaws can be propagated by planting seeds, transplanting root suckers or wild trees, grafting dormant scion wood onto actively growing rootstock, and tissue culture. The goals and purposes of the propagation effort, as well as access to resources (not all of us have access to wild pawpaw patches or, much less, sterile laboratory conditions for tissue culture), will determine how and why you are propagating the trees. Let's start by exploring starting pawpaws from seed.

Seedlings

Pawpaws can be effectively and quite easily grown from seed. Pawpaw seedlings usually represent the qualities of the parent trees to a large degree and will bear fruit in 5 to 8 years from seed, depending on care and vigor of the seedling. They do not actually "come true from seed" like tomatoes that inbreed and produce offspring nearly genetically identical to the parent plant (thus allowing seed-saving and maintenance of superior cultivars of tomato). However, if you start with excellent pawpaw genetics, the chances are high that you'll get excellent, or at least reasonably good offspring. However, wild or random seedlings vary greatly in fruit quality, and many will simply be awful and very unmarketable, so be sure any seedlings you intend to plant for fruit production are grown from very good quality parent trees. 'Sunflower' and 'Overleese' are noted as being very good parent trees to work with; excellent offspring from their respective lines is noted in chapter 11.

In summary, if you simply desire pawpaw trees of any kind, any seed will do. Wild fruits are certainly full of them. Most wild or cultivated pawpaw tree seed will sprout, produce quality rootstock, and grow vigorously when handled and planted correctly. Many pawpaw trees available in nurseries as grafted trees are simply grafted onto generic wild rootstocks. There currently are no selected or named pawpaw rootstocks, unlike apples and pears that have hundreds of rootstock options that impart dwarfing effects, disease resistance, early (precocious) bearing, etc. Someday a few choices will likely be available. However, most any pawpaw seedlings produce useable and effective rootstock that can be used for grafting. Generally, pawpaws need to be 2 to 3 years old before using them as rootstock.

Starting from Seed

So, if you have acquired some completely ripe quality pawpaw fruits and deemed their seeds worthy to plant, the process is simple, but there are some steps that must be followed carefully.

1. **Carefully wash and clean the seeds of all fruit pulp** by hand or mouth. Fruits can be masticated in water and the pulp floated off and/or fermented for a week or two to assist this process. The seeds should be very hard, smooth, dark and shiny, moist and brown-black. Make sure they are thoroughly cleaned of any pulp. Even a tiny bit left at the scar on the top of the seed may mold in storage and potentially damage or kill it. A stiff brush helps or strong jets of water from a hose work well.
2. **Sterilize seeds in a solution of either 20% bleach or 50% water and 50% hydrogen peroxide (H_2O_2) (3%).** This kills any bacteria, yeasts, and fungus present on the seeds and lengthens their viability in storage. Mix the water and H_2O_2 in a bowl or bucket and completely immerse the seeds. The solution may bubble and sizzle, which is normal. Let sit 30 minutes and agitate them a few times.

Rinse after the dip in bleach solution or hydrogen peroxide. In my experience, H_2O_2 is superior for cleaning the seeds and preventing mold and mildew. Bleach is a little too strong. Overly strong bleach solutions will damage the seed coat, making it translucent, and lower germination rates considerably, and also favor mold.

3. **Remove the seeds and place into brand-new sturdy plastic ziplock bags.** Disinfected and clean Tupperware would also likely work fine. Place a slightly moistened (but not dripping wet) folded paper towel or, better yet, a small handful of moistened sphagnum peat moss or perlite into the bag with the seeds. Paper towels work well but may start to get mildewed or moldy after a few months and pose a contamination hazard; change and refresh them after a few months. Also, this would be a good time to make sure the seeds are staying moist (not wet). Dry seeds are dead seeds. It's a good practice to check your seeds every 6 to 8 weeks to make sure they are staying moist.
4. **Label the bags with year and date and any identification information.** A permanent marker works well for this. This will help you keep track of genetics, identify different types of seeds, and also know if they have been in cold storage long enough to plant.
5. **Stratify the seeds in cold storage.** Stratify simply means to expose seeds to constant cold temperatures below 45°F, to achieve the necessary hibernation quota of the seed. Seeds were designed to stratify so that they would not fall from the tree and immediately sprout, exposing the tender seedlings to soon-arriving freezing weather. This built-in defense mechanism allows the pawpaw to propagate itself in a cold, temperate climate. In order to achieve stratification artificially, pawpaw seeds must undergo a certain amount of exposure to cold (between 35° to 40° F) storage in refrigeration. Refrigeration is the only effective and safe way to stratify the seeds because freezing temperatures will quickly destroy the seed embryo. It's quite remarkable that the pawpaw has freeze-vulnerable seeds that die when frozen and yet can still survive and reproduce in areas with temperatures that go below zero.

Somehow the seeds are insulated enough to survive this enclosed only in deer or opossum scat under a blanket of moist forest leaves and duff. Anyhow, this means that the bag of pawpaw seeds must be refrigerated for 90 to 120 days, never frozen. A brief freeze will likely not destroy the embryo but still, do not let them freeze.

6. **Once the 90- to 120-day cold-stratification period is over, the seeds are ready to plant.** The time to plant is between March and early June. This also makes sense when you are growing pawpaws in unheated greenhouses or outdoors. The freshly sprouted saplings are extremely frost-tender and will die if exposed to freezing weather or frost, and certainly do not appreciate cold temperatures. Some report seeds planted too early in the year (January or February) will sprout but then go dormant again due to the very short daylight exposure, which would not be advantageous. However, it's best to start seeds as early as is safe in your area, based on outdoor temperatures, so they can maximize their growth the first season. Of course, if you're planting indoors or in a protected heated greenhouse, you can plant anytime the stratification period is over.
7. **Seeds should be sown on their sides, ½ to 1 inch down in deep tree pots or nursery beds and covered with potting mix or very light mulch.** Sterile potting mix high in perlite and vermiculite with or without finished compost is ideal. Make sure the mix is light, very porous, very well-draining and has a light organic fertilizer mix containing phosphorous, potassium, and nitrogen. A peat moss and perlite/vermiculite blend works very well, with powdered or granulated organic fertilizer added or put on top. Pots should be at least 5 to 8 inches deep to allow some amount of taproot growth. Some growers use 12- to 18-inch or deeper pots, which work well but can be awkward to use and also heavy, making transplanting more difficult.
8. **Seeds can also be direct-sown in the ground in their final location in the orchard (or in a nursery row to be dug).** Expect a moderate success rate and plant 3 or 4 seeds per hole, thinning to

Pawpaw seeds come in a range of sizes and shapes. They can be as large as a fava bean or as small as a black bean. Better pawpaws have less seed and smaller seed (called lower seed weight).

the best-looking tree sprout that summer or next. If direct seeding in the ground, amend heavy clay soil with manure or compost (or by cover cropping) so that it is more conducive to healthy seedling growth. Overly sandy soil is not conducive, either, and needs rotted manure or organic matter added (such as compost). Heavy clay soil will hamper sprouting, as well as drought or hot sun that will surely bake the young plants dead unless covered. Covers include individual tree protector tubes, Blue-X tubes, plastic shade cloth material suspended above the rows, or double layers of polyfilm used on greenhouses and high tunnels. Clear polyfilm will increase the temperature under it so be careful if using that. Old-fashioned ginseng propagation setups would likely work as well. These consisted of wooden latticework that blocked out about 50% to 75% of the sun from the nursery beds. I heard of one grower who grew corn in his beds as shade for the nursery rows of pawpaws.

Seeds can take 6 to 12 weeks or longer to emerge above the soil. This is because the pawpaw tree has the interesting survival strategy of first securely putting down a 6 to 12 inch or longer taproot before showing any above-ground growth. This mechanism is useful because pawpaw habitat is usually very steep hilly ground or near rivers and streams. If the seeds were to not put as much emphasis on root establishment, then the typical very heavy spring rainstorm could easily wash the little seed away, as could a stream or river flooding its banks. That deep taproot usually anchors the seed in place, and the tree thus establishes safely. The fact that the little seed has enough built-in "chi/life force" to put down a foot-long root using zero photosynthesis is quite remarkable. That being said, don't expect to see the little green pawpaw shoots appear above the soil line, whether in the nursery or in the ground, for at least 6 to 8 weeks, sometimes up to 12 weeks from sowing. And, make sure to have shade cloth or protection already in place before the shoots develop, or they will sunburn and perish immediately.

It's important to note that some pawpaw seeds have taken an entire year to show above-ground growth. So, if yours have not sprouted by July or August, this may be the case. You can always try digging a few seeds up to see if they have started rooting. The bottom of plant containers may also reveal root growth (visible in the drainage holes) when there is no top growth yet, confirming that the seed is growing. Warm temperatures, regular moisture, and basic protected greenhouse conditions should preclude rapid and healthy seedling emergence, however. Usually if seeds are not showing above-ground growth within 2 to 4 months of planting, then they have died or been destroyed.

So, to repeat the most important details, pawpaw seeds must never:

- Dry out or dehydrate
- Freeze
- Be stored anywhere but a refrigerator or nonfreezing cold storage to stratify at least 90 days

Transplanting Established Trees

Can you dig up wild pawpaw trees? Technically, yes, but they should be as small as possible, and this is only effective if done in early spring, very carefully keeping roots perfectly moist and transplanting rapidly. It is not recommended and may be illegal/restricted in certain areas or private lands.

Grafting

Pawpaw trees can be grafted fairly easily. KSU has some excellent free videos online that detail the process.[1] Pawpaws take well to chip budding, whip and tongue, and modified bark grafting. T-budding gives poor results and usually fails. Pawpaws must be budding or leafing out before grafting, usually around mid-April.

A healthy strong rootstock (2 to 3 years old) is necessary for best results. A healthy strong tree (2 to 3 years old) freshly grafted should grow from 6 to 30 inches or so the first season from grafting and have close to full-size leaves (12 inches or longer). Anything less signifies either a poor graft union, plant stress or an unhealthy rootstock, or lack of nutrition.

When properly executed by a skilled grafter, a 90% to 100% success rate is common. In my observations, the better established the root system, the more growth a new scion puts on the season of grafting. It is optimal to transplant any rootstocks into pots the season before grafting, or allow one year of field growth before grafting in the field. If starting with your own seedlings to use for rootstock, plant your seeds in pots, allow them to establish at least 2 years (and build a good root system and trunk caliper suitable for grafting) and then graft, nurse one more season, and then plant in the field, making it a 3-year process from starting seedlings, grafting, and final

A freshly grafted pawpaw about 7 to 10 days after the operation shows fresh bud growth, a good sign it was successful!

This top-worked pawpaw at KSU displays impressive healthy, vigorous growth after only two seasons of growth from a tiny scion that was bark inlayed into the stump, by Neal Peterson. This tree should bear fruit in its second or third season from top-working.

planting. Transplanting rootstocks into pots and then grafting shortly after that gives less than optimal results due to the multiple sources of plant stress on the young tree.

Pawpaws are quite sensitive and slow-growing compared to other common fruit tree crops. However, once grafted, nursed properly, and planted out, they will quickly grow and can start flowering in 3 to 5 years, sometimes sooner. Never allow a newly grafted tree or one less than 4 to 5 feet tall to flower and set fruit. And, never purchase trees setting fruit immaturely below this size. This can permanently stunt them due to excess demands on the young tree's resources. Especially when commercially growing, patience will pay off in the long run, and the orchard will be more productive.

Pawpaws can also be grafted via a bark inlay graft. This is done mostly to top-work older orchard trees to convert them from one cultivar to another. For example, at KSU, Neal Peterson has been top-working old 'Sunflower' pawpaw trees to cultivars such as 'KSU Benson' and others. Those trees were around 15 to 18 years old, in good health, top-worked in early spring, and have been very successful. The successful grafts grew over 15 to 20 feet of branches within two growing seasons. This is a delicate procedure, and one should be an adept grafter before attempting this, as the entire top of the tree must be sawn off before grafting. This procedure is very useful if you want to convert some of your older trees to newer, better cultivars. Individual branches can be top-worked as well. I have attempted many times to graft seedlings or rootstock in the field via whip and tongue or the modified bark method and it has never been successful.

Pawpaw Breeding

As of 2021, very little pawpaw hybridizing and breeding work has been undertaken. Nurseryman, amateurs, and growers have undertaken some selection/breeding efforts, but so far only a few new named cultivars have been released into the nursery trade as a result (about 15 to 20 or so to date). Only one university in the US, KSU, has an active pawpaw breeding program. Thus far, pawpaw development has almost exclusively been based on locating superior wild trees, propagating these via grafting, and then starting seedlings from their fruit, which sometimes have also led to new superior cultivars. These superior wild trees and/or seedlings have also been crossbred to create hybrids. Chapter 11 lists virtually every known hybrid cultivar. Hybrids tend to be assigned numbers for identification and not names, at least not until they are released into the nursery trade. Overall, very little pawpaw hybridizing has been undertaken, or has not survived to the present day, as of this writing.

There are a few noteworthy pawpaw breeders/developers. Neal Peterson, a major pioneer in the development and propagation of better pawpaws, has worked to create a viable market for them worldwide. A big part of his effort has been in trying to find, breed, or select pawpaws that were simply high-quality enough to be marketable. Small fruits or those with poor flavor or texture or are simply too fragile to make it to market are obviously not conducive to marketing. Neither are trees that are shy producers or which get plagued with *Phyllosticta* disease, which the next chapter explores in detail. To make a long story short, once Neal became involved in trying to cultivate and locate better pawpaws, he searched the property of the Blandy

Experimental Farm in Maryland.[2] An old pawpaw planting there included some superior genetics still surviving. Although it was overgrown and dilapidated, he was able to sample some fruits and realized these were better than average pawpaw trees. From those trees, two specimens, later named 'BEF-53' and 'BEF-32', were selected. Seeds from these were grown out at Blandy. Of these seedlings, 'Susquehanna', 'Potomac', 'Rappahannock', 'Tallahatchie', and 'Wabash' were all selected. These five are generally recognized as some of the very best cultivars available today.[3] As of 2021, most of KSU's pawpaw breeding efforts are concentrated on seedlings of Neal Peterson's trees, hybridized openly in an orchard that also contains many 'Sunflower' trees and a few seedlings of Neal's cultivars. Because pawpaw seedlings turn out fairly close to the parent tree in fruit characteristics, many of them have proven to be extremely good quality, and as of 2020, two have been released ('KSU Benson' and 'KSU Chappell'). In case you're wondering, their first release, 'KSU Atwood' is actually a seedling grown from seed sent to KSU from Maryland, received in a bag of superior seeds someone sent KSU many years ago.

The late Jerry Lehman of Terre Haute, IN, was an avid pawpaw (and hardy American/Asian persimmon) hybridizer, breeder, and grower. His 40 acres of persimmon and pawpaw trees focused on trialing mostly hybrids of 'Overleese', 'Sunflower', and a few other cultivars crossed with themselves and also a triploid (hormonally altered) pawpaw. His work achieved a number of noteworthy superior pawpaws that were often top winners annually in the Ohio Pawpaw Festival. Some of these remain numbered selections. The named cultivars include 'Maria's Joy', 'Jerry's Big Girl', 'Benny's Favorite', and 'Lehman's Delight'. These have been released into the nursery trade and are considered premium, highly sought out cultivars with very large fruit and high quality. His superior selections will undoubtedly be used by pawpaw breeders in future cultivar development work. Thank you for your diligent work and useful horticultural developments, Jerry!

9

Pests, Diseases, Disorders, and Their Management

One of the challenges growers will face when contending with diseases and insects is that, as of 2021, there are very few (perhaps zero) registered pesticides and fungicides listed for use on pawpaw. So, it is currently *not legal* to spray the trees with controls, if the fruit will be marketed. I will be including OMRI-Listed Certified organic possibilities that might help in controlling these issues. These products may or may not be legally registered for use, and there is no proof they will control or minimize whatever issue is described. They are given for educational and experimental purposes only. If you are planning on marketing or selling pawpaws, it is your responsibility to contact your local Organic Certifying agency to check what products are listed for pawpaw, or contact your local Agricultural Extension Office. If you're planning on marketing the fruit, do not use products on pawpaw that are not registered for use. Let's examine the currently identified pests of pawpaws.

Insects That Feed On or Damage Pawpaw

- Pawpaw peduncle borer (*Talponia plummeriana*)
- Zebra swallowtail butterfly (*Eurytides marcellus*)
- Japanese beetle (*Popillia japonica*)
- Pawpaw leaf roller (*Choristoneura parallela*)
- Asian ambrosia beetle (*Xylosandrus crassiusculus*)
- Asimina webworm (*Ompalocera munroei*)
- Stinging rose caterpillar (*Parasa indetermina*)

- Pawpaw sphinx moth (*Dolba hyloeus*)
- Saddleback caterpillar (*Acharia stimulea*)
- Slugs and snails
- Spotted wing drosophila (*Drosophila suzukii*)
- Northern flatid planthopper (*Flatormensis proxima*)
- Brown marmorated stink bug (*Halyomorpha halys*)
- Scales
- Aphids
- Pillbugs (*Armadillidium vulgare*)
- Spider mites (*Tetranychidae* species)
- Thrips/Whiteflies (*Thysanoptera/Aleyrodidae*, respectively)
- Hornworm (*Manduca* sp.)

Animals That Feed On Pawpaw

- Raccoons
- Deer
- Opossums
- Mice/Voles
- Sapsuckers
- Domestic cats and dogs

Diseases That Affect Pawpaw

- *Phyllosticta* leaf spot
- Rhizopus rot (*Rhizopus stolonifer*)
- Blue stem disease (BSD)
- An unnamed, unidentified pink discoloration of fruit flesh
- An unnamed, unidentified foliage deformation
- Oak root fungus/*Armillaria* root rot
- Black spot (*Diplocarpon* sp.)
- An unnamed, unidentified bark lesion

Physiological Disorders of Pawpaw

- Winter bark injury
- Sunburn of fruit
- Sunburn of saplings
- Mechanical damage
- Frost damage
- Siamese fruit
- Fruit split
- Interveinal chlorosis

Insect Pests

Insect pests are usually minor on pawpaw, but can still pose significant concerns for the grower. An excellent PowerPoint presentation, created by Jeremy Lowe at KSU, features superb detailed photos of major insect contenders.[1]

Pawpaw Peduncle Borer (Talponia plummeriana)

These little guys are likely becoming the most notorious and possibly problematic pawpaw insect.[2] Previously thought to only damage the peduncle or flower parts, this insect might need to be renamed the "pawpaw borer" because it has been documented by KSU to bore through and into just about everything: flowers, fruits, twigs, and stalks.[3] Often scion wood cuttings will show its damage by a tiny discolored, circular, frass-filled tunnel in the central part of the scion wood (*frass* is a horticultural term for a sawdust-like material often consisting of powdered tree tissue and insect excrement.) The pawpaw peduncle borer starts out life as a tiny brown caterpillar, about 5 mm long, that apparently soon starts burrowing through whatever pawpaw tissue it hatches upon, causing damage to that area. Infested flowers drop or become malformed; fruit set is either prevented or ineffective due to this damage. Most years, flower damage is minimal and has been documented at about 1% to 2% of the flowers.[4] However, some years, for unknown reasons, the infestation can be severe and may damage a large percentage of the flowers. Infested trees in the nursery and/or young orchard trees may sometimes be killed by the borers. Their presence is noticeable when damaged or dead trees are cut open and the tiny brown, frass-filled tunnels are seen. Malformed flowers are

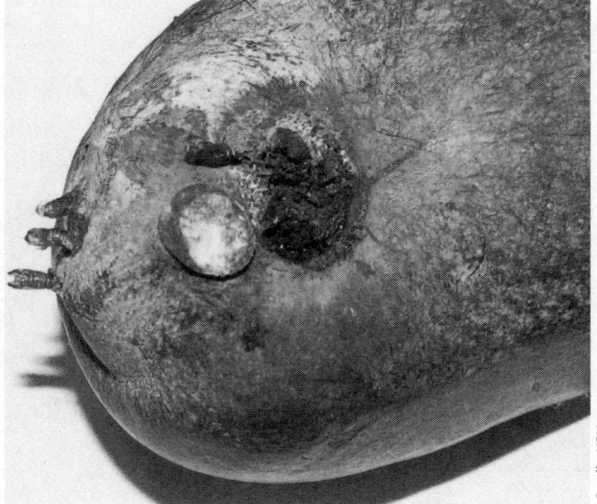

Credit: KSU

Ripe fruit damaged by pawpaw peduncle borer. Note empty cocoons on left side of fruit where borers pupated and matured, and then departed as moths.

also a telltale sign of pawpaw peduncle borer action. Fruits can also be colonized and damaged.

What might make control of this insect difficult is that it first strikes at bloom, so any spray that is applied might destroy pollinators as well, and is not recommended.

Potential control: Pyganic, although this is likely to kill pollinators as well, so discretion is highly advised.

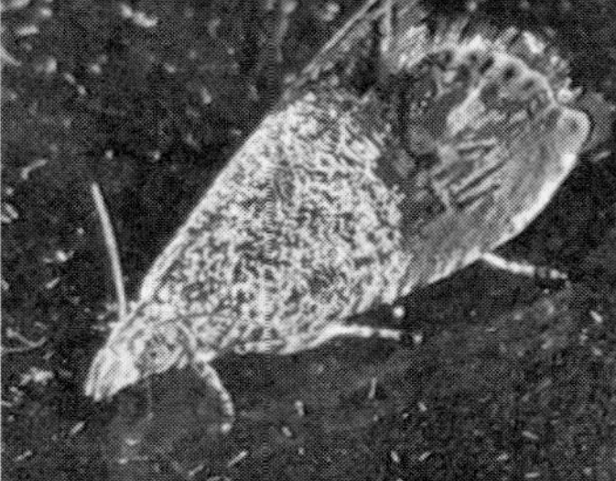

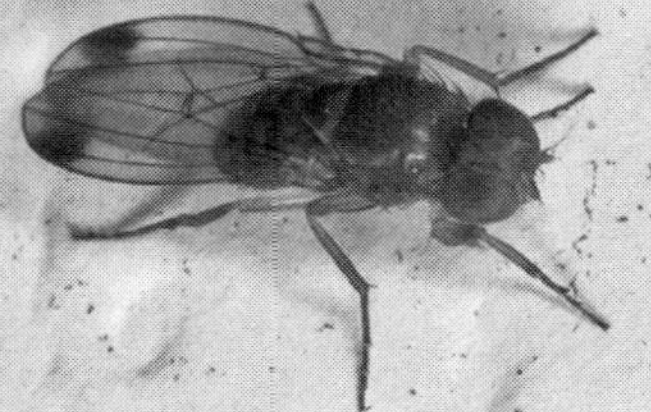

Zebra Swallowtail Butterfly **(Eurytides marcellus)**

This native beauty, known in pawpaw circles as ZSB, starts life as a tiny egg deposited singly on tops of pawpaw leaves between June and August by mature butterflies seen fluttering around the trees. *Asimina triloba* is the zebra swallowtail butterfly's only host; they have a symbiotic relationship, and the butterfly is completely dependent on them for survival of its species. The caterpillars feed on pawpaw leaves, with their appetite increasing each day as they rapidly grow. In a mature or even juvenile orchard setting, they cause very minimal damage, negligible really, but in a nursery or a new tree planting, they can cause severe damage to small trees and saplings. The trees will recover and continue growing, but obviously it does not help them when all the leaves and the tips of shoots are destroyed. Even a dozen or so caterpillars in a nursery can be pretty destructive, as a single tiny sapling pawpaw tree is like a mere snack, and many trees may be completely defoliated rather quickly. The caterpillars, eventually reaching approximately 1.5 to 2 inches long, are recognizable by their beautiful bluish-green stripes and bulging midsection hump. When handled, they exude these oddly beautiful translu-

Three pawpaw predators. *Top*: Pawpaw peduncle borer adult moth. *Middle*: The recently imported spotted wing drosophila fruit fly that invades split or damaged ripe fruit. *Bottom*: The exotic colors of the foliage-feeding stinging rose caterpillar; these will give a nasty sting if touched!
Credits: KSU (*top*); Wikipedia Commons (*middle and bottom*)

cent orange antler-like protrusions from their head (these become the antennae of the adult swallowtail butterfly).

The adult butterfly is very beautiful and large, with stripes parallel to their midsection on their yellow-silver or white-silver wings. A real native beauty. When you see these butterflies fluttering about around any pawpaw trees, either in the nursery or in the orchard, know for certain they are laying eggs. In Kentucky, they lay their eggs throughout late May to early August. Hungry larva hatch within 48 hours and immediately begin eating. Handpicking them is effective; however, you must have eagle eyes until they are at least a few days old because they are very tiny at first. Usually some damage occurs before they are of noticeable size. As these butterflies (and wild pawpaw patches) are possibly becoming more rare, it is best not to try controlling or eliminating them in a mature or juvenile orchard. What I do is carefully handpick as many as possible from our nursery trees and relocate them to a larger pawpaw tree in the orchard, or place them on pawpaws trees within a wild patch, so as not to kill them off. I have observed *Trogus penator* wasps capture and devour these caterpillars, so they do have natural predators that can also be encouraged through preserving the wasps and wasp nests. This native species deserves our protection and respect. At the same time, they must be controlled in nursery settings and brand-new plantings unless you don't mind possible severe damage. The best method is to catch as many caterpillars as possible, relocate to larger trees where they can safely feed and complete their life cycle, and eliminate the rest via organic sprays. Remember, the killing off of the larva is only temporary, for 1 to 3 seasons or so, and then your patch will be safe from damage and a literal ZSB sanctuary and breeding ground. But let your personal philosophy and inclinations guide you on this topic.

Control: Wasps are effective predators of many caterpillars, including ZSB. Try not to destroy wasps or wasp nests near orchards. Handpick the caterpillars from small trees and place them on large older pawpaw trees that will be basically unaffected by their feeding. On mature trees or healthy trees that are at least 5 to 6 years old, the caterpillars are of no real concern. If your nursery is very infested, Pyganic will kill the larva, but do try to remove and preserve as many as possible. Remember pawpaw is their only link to survival, so be conscientious.

Credits: Pawpaw Pickin's Fall 2017

Credit: Wikipedia Commons

Credit: Wikipedia Commons

Three more pawpaw predators. *Top*: Tiny Asian Ambrosia beetles showing bored tunnels in pawpaw trunk cutout. *Middle*: The lovely but sometimes troublesome Zebra Swallowtail butterfly (ZSB) adult form. *Bottom*: The familiar Japanese beetle is a mild pest of pawpaw foliage in mid-summer.

Japanese Beetle (Popillia japonica)

These notorious insects have spread across the continental US since accidentally introduced in the mid-20th century. By mid-summer, they reach plague proportions, extensively damaging many fruit and ornamental plantings. Fortunately, their presence is usually very minor on pawpaw trees. The natural compounds in their leaves may limit their attractiveness as a food source. They will sometimes cause minor foliar damage, in late June to July.

Control: Beetlejus, Venerate, milky spore (*Bacillus* sp.) via soil application, predatory nematodes, and Pyganic are all effective. Handpicking and crushing or dropping into a bucket of soapy water works also.

Pawpaw Leaf Roller (Choristoneura parallela)

The leaf roller causes damage to foliage. In the caterpillar stage, it rolls up a young leaf like a burrito and nests inside it. Adults are small moths.

Control: Possibly Pyganic or Bt products, such as Dipel.

Asian Ambrosia Beetle (Xylosandrus crassiusculus)

This newly introduced insect seems to be a fairly serious threat to pawpaws.[5] It is now commonly found on other Roseaceae fruit trees; I have seen them affect distressed fig trees as well. This very tiny beetle burrows directly into pawpaw wood, usually on the trunk, leaving a tiny "matchstick" of sawdust-like frass sticking out of the hole. Heavy infestations produce many white matchsticks, 1 to 2 inches long or so, very obviously jutting out of the trunk and/or branches. The physiology is complex: the tiny burrows they create become infested with *ambrosia* fungus that the beetles then utilize as a food source for their larvae (they're not actually eating the tree, apparently, just borrowing inside to facilitate the fungus growth). Tree death or decline often (usually) results. They seem to affect pawpaws in any stage of development that have made wood. Some growers have used termite-killing products on them. The most important thing to understand with ambrosia beetles is that they seem to (but not necessarily always) prefer and target distressed, damaged, or unhealthy trees. So the best control may be excellent hygiene, removing dead or distressed trees and broken branches, and keeping trees healthy through good nutrition and care. Keep orchards 100% free of dead trees, piles of branches, broken limbs, and other litter.

In spring 2020, we experienced a string of very late freezes after an early spring thaw and warmup that saw fruit trees and pawpaws blooming weeks early. The late freezes not only destroyed all the flowers and young shoots but seriously stressed the trees. This led to our first-ever outbreak of Asian ambrosia beetles in our orchard. Attracted to the ethylene gas no doubt being emitted by the severely stressed trees, the ambrosia beetles rapidly swooped in, and soon tiny frass tubes were seen all over many of our newly planted pawpaws (larger, 4- to 6-year-old trees showed no problems). We lost about 40 to 50 pawpaws, with 100% mortality on infested trees. We also lost many other trees, including jujubes and mulberries. So, keeping trees healthy and vigorously growing seems to be the only security measure against these insects.[6]

Control: Possibly Pyganic or organic fungus-based controls for termites. Traps made of 2-liter bottles baited with hand sanitizer or orange juice are effective. They are mainly used to detect their presence so insecticides can be applied.[7]

Asimina Webworm (Ompalocera munroei)

In caterpillar form, this builds thick webs in foliage usually near branch tips and can cause defoliation, sometimes intense damage. Old-timers used to torch webs they found in cherry trees and other fruit trees with a kerosene-soaked rag tied to a long stick. We first observed these webs in our orchard in 2019. In 2020 we had a serious outbreak, and the insects rapidly destroyed several newly planted trees before we discovered them. The telltale sign they are present is the downward-pointing dark-brown leaves the caterpillars have fashioned into a sort of "teepee" on the tips of branches. You can even see this from some distance.

Control: Possibly Pyganic or Bt products, such as Dipel. Handpicking the webs full of caterpillars and crushing them helps, as does torching. Chickens might be interested in eating them once they're grounded. This species, like ZSB, may only feed on *Asimina triloba*, so it's best to not destroy them all, in order to allow the species room to exist, but it might be necessary some years to control the population, as it was for us in 2020.

Stinging Rose Caterpillar (Parasa indetermina)

This caterpillar causes damage to foliage. It gives a nasty, painful burning sting if touched.

Control: Possibly Pyganic or Bt products, such as Dipel. Handpick (with gloves!).

Pawpaw Sphinx Moth (Dolba hyloeus)

This moth causes damage to foliage.

Control: Possibly Pyganic or Bt products, such as Dipel.

Saddleback Caterpillar (Acharia stimulea)

This caterpillar causes damage to foliage. It can deliver a nasty sting!

Control: Possibly Pyganic or Bt products, such as Dipel. Handpicking with thick gloves.

Slugs and Snails

These common garden invaders really seem to favor pawpaw foliage as a choice food source. The dimmed, humid environment of a greenhouse is an ideal environment for slugs and snails, so they can become a serious issue in a nursery if left unchecked. They also feed on pawpaw trees in the field. Damage is identified by the irregularly shaped Swiss cheese-like holes in foliage, defoliation of saplings, and dried mucus trails left on damaged leaves. They usually feed at night and can be detected with a flashlight and keen eyes.

Control: Keep grass and weeds cut short around pawpaw trees and all areas clean of debris and junk. Ducks are very effective predators. Sluggo product is very effective and is simply iron phosphate pellets coated in mollusk bait. It is OMRI listed as of 2020. Sprinkled liberally throughout the nursery or orchard, it will drastically reduce mollusk numbers practically overnight. In heavy infestations, use it every 1 to 2 weeks. It quickly molds and so needs frequent application in moist conditions. Slugs will also be attracted to and drown in jar lids or shallow dishes full of fresh beer. Ducks and geese free-ranged in a pawpaw planting for some time effectively help eliminate slugs and snails.

Spotted Wing Drosophila (SWD) **(Drosophila suzukii)**

This imported tiny fruit fly is becoming a major pest of late-season fruits, especially soft fruits like raspberries and blackberries.[8] Pawpaw's thick skin should protect it from infestation by SWD; however, in infested areas, any fallen damaged fruit or split-open fruit on the trees becomes an immediate host for SWD. Fruits infested with SWD are often unmarketable because soon many tiny fly maggots appear within the fruit. Refrigeration for 3 days kills any eggs and larva. Cooking would as well.

Control: Orchard hygiene is key. Promptly remove any fallen fruits and do not allow any to rot on the tree or ground. Maintain healthy trees and do not allow them to become water-stressed, wherein fruits might split and become a breeding ground for SWD. Pyganic, Spinosad, and other organic caterpillar sprays are sometimes effective at controlling SWD in other plantings.

Northern Flatid Planthopper **(Flatormensis proxima)**

Planthoppers cause minor foliage damage.

Control: Not known, possibly Pyganic.

Brown Marmorated Stink Bug **(Halyomorpha halys)**

These newer, very unwelcome denizens of orchards and gardens are sometimes a major pest. Not much information is available about their effect on pawpaws, but we can assume they could feed on ripe fruit. So far not much damage has been reported.

Control: Not known, but Pyganic applied to leaves and fruits would likely be effective.

Scales

These include *Lecanium* scales, tiny brown, round, hump-backed scales, and what appear to be magnolia scales (*Neolecanium cornuparvum*), a relatively large puffy white scale. Scales typically feed on trees that are stressed or receiving too much nutrition in the form of nitrogen. Ants often "farm" scales by placing them on trees and protecting them from predators in exchange for "honeydew" extract (a form of sugary exudate from the scale). If you observe ants frenetically moving about on your trees, they are probably farming honeydew, a sign of scale or aphid infestation. Scales should not be allowed to get out of control, especially on young trees, so close inspection is necessary from time to time.
Control: Horticultural oil sprayed during the dormant season will eliminate most scale issues and eggs, and is often OMRI certified for organic use. Also, make sure trees are not getting water stressed or being overly fed with nitrogen. Handpicking works for removing and destroying scales on small trees or small infestations caught early on. Ladybugs can help with scales. Pyganic and neem oil sprays would likely help. If ants and scales/aphids are a serious issue, consider wrapping a strip of wide duct tape all around the tree trunk and painting the tape with Tanglefoot, a thick, sticky, petroleum-based product that prevents ants from traveling up the trunk. It lasts one season and is OMRI-Certified for organic use.

Aphids

Aphids are tiny sucking insects, green or reddish, also farmed by ants. They will reproduce and get out of control rapidly. They are common infesting insects in greenhouses. Their characteristics are very similar to scales, except they are more mobile and reproduce via cloning, very quickly.

Control: Same as scales, above, except dormant oil is not effective against them because they are not present on dormant trees. Neem and Pyganic are effective.

Pillbugs (*aka rollie-pollies*) (Armadillidium vulgare)

Most of us are very familiar with this European import that infests rotted logs, wet lumber, and gardens. Usually harmless, nevertheless, they can take a liking to young pawpaw trees and consume the bark, causing severe damage. I've noticed them eating the bark on 1- to 2-year-old trees in moist greenhouses, especially after dark. The damage looked like little circular patches of skin missing on the young stems, and I observed rollie-pollies present. They actually breathe through gills and need a very moist environment.

Control: Hygiene and keeping areas clean and clear of rotted dead plant material and wood piles. Make sure orchards and growing areas are not kept overly moist. Handpicking and Pyganic work well. Keep weeds cut down around trees. The drier, sunnier, and more exposed to sunlight the base of pawpaw trunks is, the better it is for keeping them free of pillbugs. Of course, keep trunks painted for winter sunlight protection (see page 126).

Spider Mites

Spider mites are almost microscopic (but visible to the keen-eyed) tick-like red insects related to spiders. They make tiny webs and are usually found on the undersides of leaves, but they also eat other plant parts. Their presence is a sure sign of dehydrated and/or stressed plants. Mostly found in the greenhouse on potted plants, they thrive in low humidity indoor conditions. They cause a bleached, weakened look to leaves and can cause a lot of damage, spreading rapidly.

Control: Make sure plants are getting adequate water and airflow. An OMRI listed botanical spray, Zero Tolerance, works extremely well and is thyme and essential oil-based. Similar essential oil botanical sprays would likely work, as does soap spray. However, undersides of leaves must be sprayed, and coverage must be thorough. After the first application of an organic spray, repeat in 2 to 3 weeks and monitor. Make sure better cultural conditions are met (fans and irrigation) and keep the greenhouse clean of dead and infested plants.

Thrips/Whiteflies

Thrips and whiteflies are so similar that I am combining them. These frenetically moving tiny flying bugs feed on the undersides of leaves and thrive in the same conditions as spider mites. They cause a bleached, severely damaged, and wilted look to leaves, and spread rapidly. They can destroy or severely weaken and stunt whole blocks of trees in a greenhouse.

Control: Exact same control treatment as for spider mites, above. After the first application of an organic spray, repeat in 2 to 3 weeks and monitor.

Hornworms (Manduca sp.)

Rarely seen on pawpaws (I've only seen it once), apparently feeding on foliage. They have large appetites and eat lots of foliage, as on tomatoes.

Control: Best to likely leave them, so that braconid wasps (*Cotesia congregatus*) can parasitize them. Evidence of this is the rows of white eggs all over the backs of the caterpillars. You could handpick and put on other plants, so as not to kill them but allow them to be killed off by the wasps.

Animals That Feed on Pawpaw

Raccoons

Raccoons sometimes find their way into orchards and consume ripe fallen fruit. If you see piles of scat containing pawpaw seeds, it's likely from raccoons.

Control: Keep orchards clean and hygienic. Dogs would help keep them out of the orchard. Hiring coon hunters might be advantageous. Live trapping works, but you need big traps and it's not easy; use dry cat food and peanut butter as baits. Raccoons can carry rabies and diseases, so be careful if trapping!

Opossums

Opossums sometimes find their way into orchards and consume ripe fallen fruit.
Control: Keep orchards clean and hygienic. Dogs would help keep them out of the orchard. Live trapping works; use cat food for bait.

Deer

Pawpaw trees are generally not attractive to deer as a food source. However, being curious browsers, they absolutely will sometimes damage plantings by nipping off buds and tearing twigs off. Trees can also be severely damaged or killed by deer antler rub in winter. Deer will sometimes eat fallen pawpaw fruit.
Control: Physical barriers are the only safe, long-term way to avoid deer problems. We find it helps to use chicken wire cages around young trees and a 7-foot deer fence. Generally, deer will not go out of their way to try to damage or eat pawpaws, so even a tree tube or some chicken wire wrapped around trunks deters them easily. Dogs would help keep deer away. Hunting or hiring hunters helps reduce deer problems tremendously, and it is important for the environmental balance for us to do so on a wide scale.

Mice/Voles

Mice and voles can damage the trunks of young pawpaw trees, causing girdling and severe damage. Thick straw mulch can sometimes attract these creatures to take up residence. KSU reports that voles do not damage pawpaw trees, but some people claim they can be an issue.
Control: It's a good practice to protect any young fruit trees with a hardware-cloth tube. Simply cut hardware cloth into 8" to 12" squares, bend into a sleeve-like shape, and tie together with wire to make a closed tube. Put these firmly around the base of the tree and bury

them slightly. These also prevent string trimmer damage, which can easily girdle and kill a young tree in seconds. Cats and dogs help rid an orchard site of voles and mice. Some apple growers mulch with thick gravel or pea stone around their trees, which discourages mice and voles feeding on the trunks.

Sapsuckers (*Sphyrapicus varius, Yellow-bellied sapsucker*)

This little member of the woodpecker clan attaches to trees and punctures the bark to feed on the sweet sap within. Favorites include apple and pear, and I have found sapsucker damage on *Asimina triloba* in local parks. Damage appears as small rows of shallow holes or individual holes, cone-shaped and around ¼" wide. They don't seem to frequent pawpaws much, but this can occur on mature trees.

Control: To prevent damage, paint the trees with a latex paint tree coating, as is recommended. Wrapping the trees in burlap also can deter them. They don't seem to be of much concern, as trees that have been fed on heavily usually appear fine. However, if they feed heavily over the course of years, it could lower tree vigor and lead to decline or other pests, such as ambrosia beetle. So, if you are finding a lot of sapsucker holes, prevent more by painting the trees with a 50-50 blend of interior latex paint and water. Or, apply duct tape or tie thick burlap all around the trunk, several layers being best. Certified organic growers need to check guidelines about using paint in the orchard.

Left: Pawpaw winter damage showing crack on trunk.
Right: Typical sapsucker damage in rows on the trunk.

Domestic Cats and Dogs

It's worth mentioning that ripe fruit lying on the ground may attract pets that will consume the fruit. It seems harmless to them, but you may not want the fruit eaten or not want them eating it, for whatever reason.

Control: Keep the orchard clean of fallen fruits and keep pets well fed.

Diseases That Affect Pawpaw

Phyllosticta *Leaf Spot*

This fungus is considered one of the more serious and important pawpaw diseases.[9] It begins on foliage as tan-colored spots with dark-brown borders, which usually turn necrotic and leave a hole with a dark-brown or black halo. It also manifests as superficial (skin deep only) dark-brown to black discolored blotches on the fruit. Sometimes in mild cases, the fruit remains completely edible and even marketable, the blotches simply comparable to the black spots on a banana. However, in acute cases, the fungus may cause pawpaws on the tree to become severely blotched with black spots and unripe fruit to split open and subsequently rot. Severe infections can make the fruit very unattractive and fit only for processing (or sold only to people desperate for pawpaws).

'NC-1' is an Ontario, Canada, introduction that not only makes excellent plump fruits but is highly resistant to leaf disease like *Phyllosticta* and black spot. It is often noted as having especially lush and healthy foliage.

Control: None known but possibly monthly foliar applications of compost tea, Surround, Serenade, or other antifungal agents, such as copper or sulfur. Research into organic spray controls needs to be done based around using these products. So far that research has not been undertaken or recorded, and so any treatments would be experimental and prone to failure. However, if *Phyllosticta* is a serious problem in your orchard, experiment with one of the above treatments on some of your trees and share your results.

Rhizopus Rot (Rhizopus stolonifer)

This fungal disease, also known as black bread mold, seems to be of minor concern, primarily affecting overripe fruit in storage.[10] Fruit must be injured, either through mechanical means, rough handling, or animals in order for fungal spores to take hold. White whisker-like mold appears, and then spores develop that turn the fungus (and then the fruit) black and furry. Development is said to be rapid at 80°F and stops below 40°F.

Control: Orchard hygiene, careful handling of fruits to avoid damaging the skin, and not allowing fruits to rot in cold storage should all help. Washing fruits in H_2O_2 (hydrogen peroxide) might help prevent it in storage.

Blue Stem Disease (BSD)

This disease is still little understood and remains a minor concern in most pawpaw growing regions. However, when it appears, trees soon wilt and die when infected.[11] When dissected, the inner bark layers have a dark-blue discoloration. This disease may also cause a canker-like bark splitting on affected trees. In heavily infested orchards, newly planted trees will fail to establish and will die within one or two seasons. This disease seems to be especially virulent outside the pawpaw's native range, particularly in the Pacific Northwest. A joint USDA test planting done by several organizations in Corvallis, OR, was almost completely removed in 2002 due to severe BSD infection, but the Frankfort, KY, KSU planting is virtually unaffected. Strangely, the 'Wilson' pawpaw cultivar, although poor quality, was shown to be extremely resistant to BSD in the Corvallis trial planting, as well as the cultivar '9-58'.[12]

Control: None known, except to avoid propagating trees that are infected with BSD and be very cautious when attempting to grow pawpaw out of their native range. Keeping trees very healthy through good orchard practice; fertilizer and mulch may help them resist the disease. 'Wilson' and '9-58' pawpaw cultivars could prove to be effective in breeding resistant cultivars or rootstocks/interstems.

Pink Discoloration of Fruit Flesh

An unnamed, unidentified light-pink discoloration of fruit flesh appears to be either bacterial or fungal in origin. The affected area may

also have a whitish tinge; the rest of the fruit appears unaffected. Infected fruits could have harmful bacteria present and are not recommended to consume or use in processing.
Control: Little is known, but any fruit that appears to have this disorder should be discarded. Certain cultivars may be more prone to this, but no data is available. Preventing fruit from becoming bruised may be the only control. Don't let fruit drop to the ground and handle harvested fruit carefully.

Foliage Deformation

An unnamed, unidentified foliage deformation is also little understood and may be caused by unrelated factors present at different sites. Symptoms include a "warting" effect on leaves; twisted, extremely deformed leaves; warping and curling of leaf edges; and undersized deformed leaves. Sometimes this is present only on the tips of branches on affected trees. Ron Powell, president of the Ohio Pawpaw Growers Association, says that sometimes a tree will be affected by similar symptoms and then the next year appear normal.[13] Oftentimes such symptoms can be seen in plants of other species affected by viruses, vascular-boring insects such as leaf miners, nutrient deficiencies, or pesticide damage, such as when roadways are sprayed with herbicides and this drifts onto nearby trees. This disorder could also be a virus.
Control: None known, because these symptoms have not been identified yet. Keeping trees well fed with trace minerals by using products like kelp meal may help. Do not allow herbicide drift onto pawpaws. Tell local county workers not to spray your property and install "No Spray" signs. Do not propagate trees showing symptoms! If a tree shows symptoms for more than one year, and applying kelp and/or micronutrients does not seem to help, the tree should be removed and burned.

*Oak Root Fungus/*Armillaria *Root Rot*

A serious issue in overly wet sites, *Armillaria* root rot[14] is a fungal disease that rots out the root system of many trees, and pawpaw is susceptible.[15] However, it is apparently only an issue on sites not well-suited to pawpaw, such as overly wet low-lying areas. Symptoms include rapid dying off of trees that appeared normal the season before. Roots infected with *Armillaria* have white to yellowish fan-shaped mycelium between the bark and the wood. Dark-brown to black structures that resemble shoestrings sometimes can be seen on the root surface. At times, large densely packed, honey-colored mushrooms form at the base of infected trees in late fall/early winter after rains. **Control:** The only form of control is prevention. Only plant pawpaws on well-draining sites and avoid over-watering with irrigation. Remove any trees showing symptoms. Once trees show symptoms they cannot be saved.

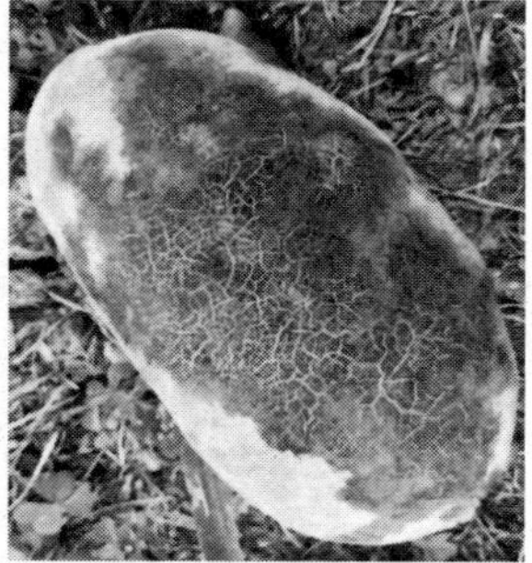

Credit: Blake Cothron

Black Spot (Diplocarpon sp.)

This harmful fungal disease has just recently been identified, thanks to the help of Ron Powell.[16] In very wet and rainy seasons, this disease appears as blotchy black spots on both sides of the leaves. The undersides will have a patchwork of black spots. The leaves gradually turn yellow, droop, and often drop prematurely in August or September, thus exposing the fruit to sun scald. This lowers tree vigor if it continues year in and year out, and can eventually (and usually does) kill the tree. It does not appear to infect the fruit, but can negatively affect it indirectly through loss of foliage and shade. Some cultivars show some resistance to the disease, but it is too early to name any resistant cultivars for sure. We have identified this disease in a few of our trees in our orchard in KY. It usually begins to appear around June when the leaves are full-sized. Organic antifungal sprays and compost tea (sprayed on foliage) may help, as well as increasing air-

Top: Pawpaw fruit showing an unsightly, moderately severe *Phylosticta* infection. Severe infections often split the fruit. *Middle*: Black spot (Diplocarpon Sp.) infection on foliage. Note bright yellow halos around necrotic black lesions. This is a serious disease that can kill entire trees. *Bottom*: An unidentified bark lesion on a young tree. Note the ropy inner bark made visible by the peeling, necrotic cambium, as well as the dried out sapwood of the trunk.

flow in the orchard, pruning excess branches, and keeping trees very healthy and vigorous. With climate change, it seems springs/summers in KY and other areas are tending toward increased rains, so expect this disease to become more of a concern in the future, and avoid planting highly susceptible cultivars.

Unidentified Bark Lesion

I am calling this simply "pawpaw bark lesion." It is not known what causes it. It could easily be confused with mechanical damage, but I have noticed it in my grove where no mechanical damage occurred. It can appear on the twigs as well as the trunk. It looks like something ripped or sanded the tender outer bark, exposing the inner ropy bark. It may be the same condition as BSD, and is certainly harmful, but its identity is not known at this time.

Physiological Disorders of Pawpaw

Winter Bark Injury

Also called "southwest injury", because it commonly occurs on the southwest side of the tree that gets the most direct sun exposure in winter. This common cause of fruit tree damage and death *must* be prevented in a pawpaw grove. Basically, it is caused by the freezing of trunk tissue, which, on warm days or when exposed to strong direct winter sun, subsequently thaws out temporarily, expands, and then is quickly frozen at night, thus shrinking and rupturing the tissue. This often leads to very large open wounds spanning the entire length of the exposed trunk.

Top: Severe trunk split caused by winter sun on the south side of this pawpaw trunk. This severely damages a tree but may not necessarily kill it, and the trees can sometimes recover if subsequently painted. *Bottom*: A nicely protected tree at KSU showing thick latex paint application and no winter trunk damage.

This grafted pawpaw tree is in severe decline after about 20 years in the ground and minimal care the last 5 to 10 years or so. Note the large split in the trunk from winter sun and small yield of fruits underneath. Any fruit produced are highly prone to sunburn. Heavy pruning and heavy fertilization, as well as painting the trunk, could probably give a tree like this a few more years of some fruit production.

The subsequent large open wounds on the trunk greatly diminish tree vigor and sap flow and will reduce fruit yields, and may cause a tree to perish or go into decline. Look for long vertical tears on the trunk.

Control: KSU recommends whitewashing the trunk of all pawpaw trees with white paint, which reflects the winter sun and completely prevents damage from occurring. A typical fruit tree whitewash is a solution of equal parts interior latex paint and water, thoroughly mixed and painted liberally on the trunk and undersides of the branches, and reapplied annually. Many growers say to avoid using outdoor latex paint as this has additives that may damage fruit trees. Certified organic growers will need to check regulations around using latex paint in the orchard; it is usually not allowed. Clay and cow manure mixtures also work but must be reapplied more often. A recipe for a natural sun protection formula is equal parts cow dung and native clay (cow dung can be replaced with humus/compost and sand). If your clay is dark, add milk powder to lighten it. Applying a heavy coating of Surround would likely work but may need several applications each winter. Wrapping the trunks in burlap would also help, and the burlap should last a few seasons.

Sunburn of Fruit

Sunburn occurs when defoliation, poor tree health, storm damage, pruning, or some other factor causes a reduction in foliage cover on the tree. Inferior cultivars could also have less foliage cover and subsequent exposure of fruit

to sun damage. Exposed fruit is often exposed to direct afternoon sun where it essentially bakes. The skin on burned fruits turns bright yellow and appears bleached. Fruit quality suffers, rendering them inedible; they do not usually ripen properly. Fruit sunburn can occur in late summer or early autumn as trees are ripening fruit. At this time, the trees are sometimes beginning to lose their leaves, yet the sun is still hot and intense and can now sometimes directly hit the fruit due to the more direct angle of the sun at that time.
Control: Keep trees healthy and vigorous by fertilizing with nitrogen 2 or 3 times every spring and irrigating as necessary. Avoid pruning pawpaw trees except to reduce height and remove broken branches. Prune only when trees are dormant so as not to expose the fruits to direct sun. Surround spray may help reduce sunburn when used on exposed fruits.

Sunburn of Saplings

All pawpaw seedlings need direct sunlight/UV protection until they reach around 30" to 36" tall. Grafted trees under 18" require UV protection as well. Seedlings that do not receive UV protection become damaged, stunted, and usually die.[17] Sunburned trees will have a bleached look to the leaves, often with brown lesions; they will be small and soon wilt, and the tree shortly dies. A sunburned tree can possibly recover if it is provided sun protection immediately after getting burned, although growth will be reduced that season.
Control: See chapter 4 for complete details on tree UV protection.

Mechanical Damage

Mechanical damage is completely avoidable in most situations. This includes string trimmer damage to bark, lawnmower or bush hog damage to exposed roots or the trunk, and soil cultivation/tilling that damages roots. Winter bark injury can be confused for mechanical injury, and vice versa. However, winter bark injury only occurs on the exposed south- and southwest-facing sides of the trunk and usually spans the whole trunk vertically on that side. Mechanical damage affects the bottom 6 to 8 inches or so of the tree, usually near the surface of the soil, and the bark appears flayed, split, and jagged, with the inner bark exposed and dried out.

Control: Keep trimmers and machines a safe distance away! Keep trees heavily mulched at all times so that string trimmers and close mowing are not needed. Use flame weeders (on larger trees only; flaming will damage leaves and young stems and should not touch the trunk of older trees!). Put hardware-cloth tubes around trees to prevent string trimmer damage, especially if you hire people to work in the orchard. Zero protection on the trunk will almost certainly lead to eventual damage, if not by machines then by voles and mice. Protect the trunks with paint and barriers.

Frost Damage

Frost damage is not unusual, as pawpaws flower rather early in the spring, and young leaves soon sprout after that, generally all at that magical, life-infused yet risky time where fluctuating weather can lead to a damaging spring frost. Light frost damage may brown shoots and young leaves and damage them slightly. When frost damage is severe, leaves, young shoots, and flowers turn black, appear burnt, and fall off. Severe freeze damage after growth has commenced may lead to the graft portion dying on a very young tree, with only the rootstock sprouting back. Usually trees bounce back quickly from frost damage, but they may not grow quite as much that season and, depending on severity during flowering, may not set a crop, or only a reduced crop. The good news is that pawpaw flower buds are very cold resistant and have such a long flowering cycle (spanning 2 to 3 months) that frost events usually do not eliminate a crop from forming and, in many years, do little to no damage.

In early May 2017 in central KY, we experienced a light, unusually late frost at 37°F, two nights in a row. The first one caught us off guard and caused minor damage to some freshly planted very small grafted trees. Those most adversely affected lost their terminal buds and had to resume growing from lower buds. This reduces the new growth a tree puts on in a season. Now we were expecting frost, so the second night, we completely and very carefully covered the newly planted 1- to 2-foot-tall trees with thick fresh straw, and they did not sustain further damage. The trees recovered and did well that season. Open full blooms and young fruitlets can also be destroyed by frost, and will subsequently fall off the tree. Pawpaw peduncle borer damage could be mistaken for frost damage to blooms, as they both make blooms shrivel. However, frost damage only happens the night after a frost event.

Control: Kelp extract applied as a foliar spray is said to help trees handle cold stress. Old-fashioned kerosene orchard heating lamps might be a good investment, but these are rare and expensive. Used ones can be found for sale online and in citrus regions such as Florida. Modern propane orchard heaters are also available. Bonfires, large fans, as well as misting systems are all effective for other fruit growers when minor frosts threaten. Covering small trees with woven geotextile row cover fabric, straw, blankets, large tubs, 5-gallon buckets, barrels, 55-gallon drums, or even trash cans helps prevent damage when temperatures dip into the 30s after growth has resumed. Placing cement blocks or weights on the drums will prevent heavy winds from tossing them over. Avoid planting pawpaws in low spots and frost pockets. Late-blooming cultivars may be desirable in late frost-prone areas. So far, late-blooming cultivars have not yet been identified. However, pawpaw has such a long flowering period that even if some blooms were to be destroyed in a single late frost event, a harvest would still be obtained most seasons, albeit diminished.

Siamese Fruit

Imagine two pawpaws that fused together at the seam, and you'll get a picture of "Siamese fruit." It is still edible and useable but looks strange and is very fragile, as the seam can easily rip, exposing the flesh of both fruits to the open air. There are theories about why this happens but nothing conclusive. It's common to see one or two Siamese fruits in each tree every season. It is harmless, more an anomaly than anything.

Severely split pawpaw due most likely to physiological reasons and not *Phyllosticta*.

Fruit Splitting

Various factors can lead to fruit splitting. *Phyllosticta* infection is the most likely culprit, as is overwatering after a very dry period. However, sometimes fruits simply split open when nearing maturity even when not affected by *Phyllosticta* (and have no sign of black blemishes or spots common with that affliction). The exposed flesh often seals over, and the fruits remain marginally edible but not sellable. Seeds from such fruits should be discarded as they usually become damaged by desiccation. Do not let your orchard dry out severely during drought and then attempt to remedy with heavy watering. This dehydration followed by inundation with moisture can cause fruits to split. This happens with many other soft fruits such as cherries. Make sure orchard soil moisture levels remain fairly consistent through monitoring, with deep and regular irrigation if there is a drought, slowly over the course of a week or so. Aim to keep the orchard constantly at an even, slightly moist level with plenty of mulch and irrigation as needed.

Interveinal Chlorosis

This disorder is believed to result from a lack of magnesium and/or potassium in the soil. Neal Peterson says pawpaws need a lot of potassium/potash in the soil. Symptoms are a bright yellow coloration between the veins on the leaves, creating a fractal-like pattern with the green veins. After treating, the symptoms may not disappear that season but should not reappear next season. To treat, apply to each affected tree a few tablespoons of Epsom salts and a source of potassium (greensand, wood ashes, etc.). Also, kelp extract sprays can be used. Large trees may need higher amounts of these nutrients, but no information is available right now. The best thing is to make sure all trees are well-fed with compost, manure, fish emulsion, kelp, and thick mulches in order to avoid nutrient deficiencies.

10

Pawpaw Fruit Marketing Strategies

So, what do we do now with a hundred or even thousands of pounds of soft, dripping ripe pawpaw fruit? This chapter will help you avoid that question! Like any commercial orchard endeavor, you absolutely must establish the fruit distribution plans months, if not years, before the fruit ripens. Otherwise you may find your profit margins and pride hurting badly as fruit rots in storage or in the field.

The best option for the market grower may be to combine three or more of the following strategies. This is very important in case one avenue of distribution has a bad season or the deal falls through, you have some backup markets for moving your product. Always have a plan B and C when marketing fresh produce!

The main methods of selling pawpaw fruits are as follows:

- Fresh fruit sales at farmer's markets
- Fresh fruit sales to grocery stores
- Fresh fruit sales via mail
- Fresh fruit sales at U-Pick farms
- Fresh fruit sales to restaurants and breweries
- Freezing fruit to sell as frozen pulp
- Value-added goods
- Ethnic markets

Fresh Fruit Sales at Farmer's Markets

This is oftentimes one of the least important avenues when trying to sell pawpaws because they are still an unusual, often unknown, and thus mostly undesirable fruit, compared to the strong appeal and easy market acceptance of peaches, apples, and the like. However, some people, including Neal Peterson, did do this successfully by creating hype, offering free taste samples, and making this his main product at a farmer's market in Washington DC, where their large farmer's markets may dwarf your local ones in size and customer base.

In 2020, fresh pawpaw fruit sold from $1 to $5 per pound. I personally would definitely charge more than $1 for retail sales. Based on extensive farmer's market experience (although with other products), I would not expect fresh pawpaws at an average farmer's market in the US to bring in more than $50 to $200 per week, roughly 25 to 100 pounds, at $2 per pound. That may even be optimistic, depending on the size of your markets. With the perishability of the fruits picked throughout the week, a large yield from a grove would mostly go to waste if a farmer's market was your only outlet—not the way to do it.

In order to sell pawpaw effectively at a market, design and purchase handmade, high-quality vinyl banners from local sign makers to advertise the fruits via bright colorful pictures, large words (PAWPAW FRUIT!), and basic information. The advantages to vinyl signs are they are very light, can be rolled up, and are durable. Create lots of hype by informing your customer base weeks ahead of time of ripening that you'll be bringing pawpaws to the market. Send out marketing emails, social media posts, etc. weeks before ripening and a few days before taking the fruit to market. Inquire if your local farmer's market coordinators can help create some hype by messaging the customer base. Call the local news stations and newspapers weeks before they ripen and create as much fervor as possible. You are doing this ultimately to sell the pawpaws right?

At your table (and in the field) make sure all fruits are picked and handled delicately, cleaned, and stacked very attractively like at grocery stores (but not so high that they risk getting crushed or spilled everywhere). A key to being successful at a farmer's market is to make your table look as attractive, stacked, abundant, active, and inviting as possible. Always be extremely friendly and engaging at all times (never sitting down or playing on your phone!). Don't have your fruits lying scattered around your table with a little paper or cardboard sign scribbled with a price. That looks messy and uninspiring and will lower appeal drastically. Put them in attractive wooden display boxes lined with burlap or cloth, and stack them high. Remove any split or broken fruits. By all means, offer free samples or free fruits (always abide with local food processing laws if cutting open fruits for sampling). Create quality laminated signs with the names of varieties (or simply Fresh Pawpaw Fruit) and the prices. Divide them by cultivar if possible, with descriptions. Give out recipe ideas (ice cream, pawpaw bread, jams, sauces, etc.). Be ready to answer lots of questions and be super friendly. Maybe create an informative handout. Have some bulk prices ready in case someone happens to want 10 or 20 pounds, or bruised fruit for discount. Handle the fruit carefully at all times, and politely make sure your customers are not groping and fondling them too much (PLEASE DON'T SQUEEZE THE FRUIT signs, perhaps?). Bag them carefully and don't stuff too many into a sack so as not to mush them. Maybe use paper bags, which will slow the ripening process, compared to plastics. If someone just spent their hard-earned money on an already unusual product, just to get home and find them mushed and oozing everywhere, they will not buy again. Bring some shallow cardboard boxes so people can take home a lot if they want, safely.

Overall, consider pawpaws to likely become a small supplement to your existing farmer's market sales, not as a major outlet for the fruits. Markets in large cities will fare better than rural ones, if you create enough hype and get known for having great pawpaw fruit. That being said, many rural markets in the South and also the lower portions of Indiana/Ohio will likely have some older customers who suddenly might become very sentimental about pawpaws and excited when you have them (or better yet, tell them weeks ahead they are coming soon). However, getting them to buy more than a few or to pay $3 per pound may prove challenging.

Fresh Fruit Sales to Grocery Stores

This may be the best route of all, but will take some tact and strategy. First, the commercial mega-grocers are going to have little to zero interest in entertaining your pawpaw fruit sales pitch, unfortunately. That may change in time. If somehow they do, they will usually only want huge amounts of perfect fruit at low prices, and if you don't deliver, the door will shut and not reopen. They may also want you to have some kind of insurance or GAP certification, so be informed, prepared, and upfront.

However, the small to medium-sized gourmet/local food/health food stores can be excellent places to distribute fruit. In Kentucky, there is a local medium-sized modern health food grocery store that annually purchases about 400 to 500 pounds of fruit from a local pawpaw grower. That's impressive! It then retails them for around $2 to $3 per pound. Small stores may only be willing to handle 50 to 100 pounds, but that's an easy and quick way to move some fruit. Discuss this with the storeowners many months in advance of having ripe fruit, give them exact dates when you think it will be ready, and stay in contact when the picking season is nearing. Make sure the fruits you deliver to the store are perfect, undamaged, good size, and sparklingly clean. Only sell firm-ripe fruits—no fallen, bruised, or overly ripe fruits! If the fruits are delivered overly ripe or bruised and begin to spoil in a day or two while on display, the store manager will be displeased. Let them know what to expect, that the fruits bruise easily and usually need refrigeration, and give them ideas on how to hype the fruit via social media, newspapers, etc. Make sure you're exactly on time with everything and very communicative if there are any issues (for example, later or earlier ripening than expected, crop loss). Decide on a fair price before you deliver. A good wholesale price would be $1 to $2 when moving that much product. Remember, the store has to increase that to a retail that their customers will actually pay. Don't expect to make $5 a pound when moving bulk amounts! However, the ease and convenience (and success) of selling in bulk makes up for the lower price received.

When trying to find retail outlets to buy bulk amounts of your fruit, contact as many places as possible, months in advance, and be persistent. Don't leave messages; simply talk to the managers directly or send a detailed email and follow up with phone calls. Go in person to discuss the details. Expect maybe one in four or five places to be interested. Explain how pawpaws are becoming very popular with gourmet cooks and send some links to websites if they aren't sure what they are. You might just seal a sweet deal and be moving hundreds or thousands of pounds this way. Just be persuasive, polite, persistent, and punctual! The four P's we could say.

Fresh Fruit Sales Via Mail

This newer avenue for selling fruit takes lots of extra effort and marketing, but you can usually get the best prices. With the extreme popularity of mail-ordering online these days, thousands of farmers are now selling products, including fresh raw produce via websites, online food distribution companies, and catalogs. Raw foodists, vegans, and the Paleo crowd can all be likely target customer bases. This marketing approach can be very effective, but takes some marketing savvy and perfect follow-through.

First, you must create an online presence through a modern, attractive and easy-to-use retail marketing website (built primarily for mobile phone use) and/or get listed on some of the many local food/fresh produce marketing websites becoming popular right now. This is beyond the scope of this book, but web developers, websites, and books on this topic are available. Consider hiring someone (read: hire a millennial) to develop this for you. Basic ecommerce websites can be developed for about $500 to $2,000, depending on complexity and size. Likewise, a much easier, cheaper, and possibly more effective approach would be to create listings on various websites that advertise and distribute local foods (Etsy-style). Also you can market products at retail prices directly through Amazon or eBay. Investigate.

Unless the website you are using requires certain shippers/carriers, you may choose to utilize the United States Postal Service (USPS),

United Parcel Service, or Federal Express (FedEx). Each has their pros and cons, including pricing, availability of services in your area, and package handling. Choose carefully considering all those factors. Just for the record, we almost only ever use USPS to ship our products.

Whatever the case, be extremely exact on what you are selling, what customers will be sent, when they can expect delivery; provide lots of clear, attractive photos and details. People love lots of details and pictures these days! One word of caution: be careful about giving customers too many options as it can intimidate them or cause you a lot of headaches (for example, promising certain cultivars of pawpaws). Make sure to keep your listings updated, detailed, and accurate so people are not trying to order fresh pawpaws in February or June. Make it clear they only ship from August through October, or whenever you have fresh fruit.

You must sell only perfectly unblemished, flawless fresh-harvested fruits (shipped the same day of picking), shipping orders overnight express Monday through Thursday. Orders shipped on Fridays may take till Monday to arrive, and will likely be a fermenting mess upon delivery. If your customers receive a box of fermenting pawpaw mush, they could get very upset. Replace any messed-up orders immediately or refund. These days if you're using a retail website platform such as eBay or others to market products and they arrive in poor or spoiled shape, the website will refund the customer and you may get bad ratings. Therefore, package everything like it's made of thin glass! Use lots of insulating material and make sure the package is attractive and presentable (not dirty or covered in random tape, etc.). Styrofoam is very bad for the environment and ends up in the dump or as litter, so look into other options for packing materials including biodegradable Styrofoam-like peanuts or inflatable bubble-type wraps. Thick newspapers or craft paper usually works quite well. If you shake the box and can hear or feel anything moving, it's not packaged properly, and the contents will arrive damaged and your customers will be peeved. Open it and repackage with more insulating material. The heavier the box, the harder it is to get it in great shape to someone's door. So, you may set a reasonable limit (5- or 10-pound order maximum). A general rule is that no two fruits should be touching or pressed together without packing material snugly in between them and the walls of the box. Practice packaging fruit and mail it to a friend, yourself, or a family member and see how it arrives. Take notes on what worked or what didn't, and alter accordingly. Include a copy of their invoice and your business card (you have one, right?).

Mailing fruit can be effective and profitable because you can charge very high prices, and the entire country becomes your market, not just your local area. Mail-order fruit should go for $5 to $10 per pound. But remember, if you get lots of upset customers who received mushy fruit, your business will dry up fast. A few negative or nasty online reviews can hurt your image quickly. Request your happy customers leave great reviews for you. This looks good and gives new customers more confidence, and it buffers the negative effects of any bad reviews you may receive.

In my opinion, shipping pawpaw fruit across the country is risky business, creates a lot of waste, and has a high carbon footprint. Sell the fruit locally if possible; it's a lot easier, faster, less risky, and the greener approach. However, I include it because it may be your only or best option.

Fresh Fruit Sales at U-Pick Farms

Once again, this is likely to be a very minor income stream, but it can become a niche crop to create more revenue and excitement at your existing fruit farm. Because pawpaw harvest season coincides nicely with most apples and pears, they could easily become another fresh fruit product offering available for customers already walking onto your operation. You will probably prefer to harvest the fruit yourself and sell it from a storefront instead of having people attempt to harvest it themselves. It's likely they might damage the trees or harvest hard immature fruit that will never ripen (and cause upset and disappointment). You might even hang signs around the trees saying as much. In addition to fresh fruit sales, your U-Pick could offer pawpaw bread, jams, ice cream, etc.

If you live in an area where pawpaws are already popular or known, you may get some good sales from adding a grove of 30 to 50 pawpaw trees, depending on the size of your operation. They could also be sold via your existing networks, such as local grocers and farmer's markets. Make sure to advertise them heavily so people know you have them, and once again, build lots of hype weeks before they are ripe.

Fresh Fruit Sales to Restaurants and Breweries

This is sometimes the most effective way of moving lots of pawpaws quickly and easily. First, we'll discuss restaurants because the two avenues operate much differently.

Restaurants are usually somewhat difficult to work with; just ask any long-time local produce farmer. The reasons for this are that restaurants have a high turnover rate of employees, including chefs. These days chefs can sometimes get a bit of a "rock star ego" because of all the foodie hype and hit TV shows featuring chefs, etc., which makes them even more difficult to work with. In my experience selling produce to restaurants, I observed that chefs sometimes get very excited hearing about your local gourmet products. They then often start spinning off ideas of everything they could do with your products; giving you a very positive impression they're going to become a big buyer. So, you make an agreement to bring in X amount of Y (such as 50 pounds of pawpaws). You then do all the hard work, and months later, at the agreed-upon time, you contact the restaurant, only to find that the inspired chef has moved on and the restaurant has no idea what you're talking about, or that things changed and they only need 20 pounds now (not 50), or that they will take the 50 pounds but then complain about the quality (such as seediness of the fruit, the short shelf life). To be fair, many chefs and restaurants are reliable, reasonable, and easy to work with. However, be very careful about getting your hopes high when a chef acts very excited or makes promises and plan wisely. Check in several times with the chef over the season to make sure you're on the same page.

To go about this properly, you'll have to do some homework and discern what restaurants would be interested in pawpaws. Obviously, large commercial and corporate restaurants will have zero interest; these types of businesses usually don't purchase local produce anyway. Make a list of all the local restaurants you'd be willing to deliver to, including bakeries, ice cream shops, and coffee shops. Then contact all of them and ask to speak to the chef or owner. You likely won't reach them immediately, so don't leave a message; keep calling back every couple of days till you reach them. Create an organized list of the names of the owners, chefs, and the restaurants. Ask to have a little meeting to discuss your local farm and what you can provide them. When it's time to meet, dress neatly and have a professional attitude like you're going to a job interview. Be as organized as possible, bringing some photos, business cards, and even some fresh produce to show off in a clean cooler. Have a notebook to take notes. Provide a list of your products without prices. Tell them a brief story about your operation and products and when they are generally available. If other restaurants are already clients, make sure to mention them and the chef's name. Never discuss prices unless they bring it up. If they ask, say you'll email a list of all your prices and do so the next day. If you hear nothing back, contact the chef to discuss them. Prices are touchy because if they are too high, the chef will quickly forget about you and move on. Expect your pawpaws to sell for about $1 to $3 per pound. You may have to haggle on prices, but make sure the price is fair and sustainable for your operation. Remember, even if they only want to pay $1 a pound, because they're likely buying in bulk, this is very easy and convenient for you, and that's part of the trade-off.

If you make an agreement with the chef, remember it's not written in stone. However, act like it is: check back a month before fruits are ripe to say they'll be ready soon and make sure he's still on board.

Then contact him a week to ten days before harvesting and schedule the best time to deliver them, making sure you fulfill that flawlessly. If you are a day late, the chef could become extremely upset and may not do business with you again. Chefs develop their menus weeks or months in advance and may be planning on making a special dish out of your pawpaws for a big Saturday event, or the like, and if you are late and they cannot make the dish, this will make them very irate and might disappoint their clients. No more business for you. So, be on time perfectly. If somehow something crucial comes up, deliver them a few days early and apologize and explain why. That is much better than delivering late, and the chef will be happy he got the goods. Box up your delivery nicely, preferably in waxed produce boxes, but not so high that fruits get bruised or mushed. Make sure all your products are sparkling clean! Restaurants will be very displeased if your produce arrives dirty. It must also be excellent quality, with no tiny or low-quality fruits tucked in. If you get a good reliable chef, you may be able to move a lot of fruit at their restaurant. And, you may choose to grow additional products for them, such as vegetables or other fruits. Just keep in mind that if the chef leaves the restaurant, the next chef may not be so easy or reliable to work with. That's why you'll want to contact as many restaurants as possible. Remember, plans B and C.

Breweries are a bit different and might prove easier to work with. Pawpaw beer is becoming something of a hip gourmet item in some areas currently. Contact local breweries and wineries and talk to the owners at least 3 to 4 months before fruit is expected. Try to set something up, on the premise you're a local farmer and can provide interesting pawpaws later in the season. Once again, go as though it's a job interview and be very communicative. If they act favorable, try to get them to agree on a certain amount. This may take a few weeks because they may have to do some research and figure out how much they'll need. But follow up with them and use the same rules just discussed for restaurants. Breweries are a little more consistent, and sometimes want huge amounts of fruit. I know of one grower who sells over a

1,000 pounds a season to a local brewery in Ohio. Once again, just be communicative, consistent and follow through with a clean, high-quality, and timely delivery. Once you're in, you may be able to provide other local products to the brewery as well. And, in all cases, if you are unsure you can provide a large order, such as 1,000 pounds of pawpaws, be up-front and give a safe estimate of the amount and stick to it. If you'll have less than 50 to 100 pounds, a brewery is not the place to try to sell. However, some brewers may want to experiment brewing with pawpaw the first season, and may only want to start with a trial batch of 50 pounds or so. But try to make sure that goes well by providing high-quality fruit on time, and next year they may want 1,000 pounds. Remember, even though they're fermenting and brewing the fruit, that does not mean it's OK to get them bruised low-quality fruit. You may decide to sell a brewery your less retail marketable B grade fruit of smaller size and lesser quality, but make sure it's still sound, unblemished, unbruised, fresh, and good quality.

Freezing Fruit to Sell as Frozen Pulp

A few growers have found this to be a good niche for them. Basically, huge amounts of fresh pawpaws are pulped and deseeded, mostly by hand, bagged, and immediately frozen. KSU and the OPGA provide some information on doing this with a food processor designed for processing tomatoes. It works with a little simple modification and can be found online.[1]

Any pawpaws used for this purpose should be clean, unblemished, fully soft-ripe, fresh, and not have any bruising or pink discoloration. All seeds must be removed. The extracted pulp must be frozen immediately after being bagged, or it will quickly ferment and spoil. Before adding pulp or freezing, attach bulk printed attractive stickers to the bags. Make sure the bags stay clean and have a consistent weight. Advertising online will help move this product. Restaurants, breweries, and bakeries may be interested in frozen pulp, as it would extend the pawpaw season indefinitely.

Be very aware that in order to process produce and sell value-added (processed) products, you must contact your local USDA branch and discuss all their rules and regulations. Ignoring the laws, which vary tremendously from state to state, can result in huge fines or a shutdown. Sometimes a product will have to be labeled something to the effect of "Processed in a Home-based Facility." There are rules about cleanliness, and you may need to be inspected. If you desire to sell outside your state, you will need to build a commercial kitchen, which requires electricity, running water, a central drain, and a pet-free area that is kept clean. This can be built for less than $10,000 if you do it correctly and small-scale. Grants may be available. Your local agricultural extension office can direct you to programs and local regulations.

Value-added Goods

Pawpaw pulp can also be made into jams, wines, beers, breads, muffins, pies, etc., all marketable products. The nice little secret is that, like many fruit growers, your A grade premium fruits can be sold retail/wholesale, while your B grade fruits can be processed into value-added goods. It's a great use of good-quality yet small or unmarketable fruit. It also can make the operation more profitable by using the fruit constructively.

Experiment with various recipes and get a taste-testing group together once you are pleased with a recipe. When you have a good recipe down, try selling the product. Farmer's markets can be good places to move value-added goods. Other possible products include: pawpaw hot sauce, chutney, herbal products (such as lice shampoo, pills), body lotions, and jewelry made of pawpaw seeds. Make sure to follow the same advice above and check local regulations first and obey all the laws. Do not neglect the power of Internet sales and marketing with value-added goods. These generally are small, shelf-stable, and easy to ship.

A great place to start with value-added goods is a book called *The Edible Pawpaw*. See recommended reading for full details. If you're serious about getting into value-added goods, you need to read this book. Although it contains hundreds of recipes, not all are tested, so be careful when trying.

These are some pawpaw products that are being currently sold successfully:

- Ice cream
- Hot Sauce
- Jam
- Wine
- Beer
- Chutney
- Frozen pulp
- Baked goods (locally)
- Health food supplements (pills containing extracts, etc.)
- Jewelry made out of pawpaw seeds

Other pawpaw products could include: smoothies at major smoothie outlets or trendy raw food restaurants, cheesecake, face masks for luxury spa centers, lotions, lassi at Indian restaurants, barbecue sauces, salad dressings, vinegars, and...? Get creative! Any recipes that include bananas could use pawpaw instead. Following are a few recipes (slightly edited) taken from *The Edible Pawpaw*.

Pawpaw Lassi

(Indian-style milkshake)

This recipe was created by chefs at Casa Nueva Restaurant in Athens, OH. The winner of Best Drink at the 1999 Festival.

½ cup water
4 cups plain yogurt (Greek style might be even better)
⅔ cup raw honey
½ tsp. sea salt
1 cup pawpaw pulp
½ tsp. ground cinnamon

Blend until smooth. Makes 3 to 6 servings.

Sweet Hot 'n Tangy Barbeque Sauce

(submitted by Al Schmidt)

1 head garlic, roasted
1 onion, roasted
1 large tomato
1 apple (Granny Smith or something sour might do well)
1 jalapeno pepper
2 long green hot peppers
½ cup apple cider vinegar
2 tsp. non-GMO cane sugar
½ tsp. paprika
½ tsp. Worcestershire sauce
¼ tsp. achiote
1 cup pawpaw pulp
1 onion, minced and sautéed in 3 tbsp. oil and 2 tbsp. butter

Puree all ingredients except pawpaw. Stir in pawpaw pulp. Simmer for 5 minutes.

Vegan Pawpaw Zucchini Bread

Submitted by Amanda Shawie and placed 2nd in the bread category at the 2006 Ohio Pawpaw Festival.

5 cups wheat flour
1 cup shredded zucchini
¾ cup pawpaw pulp with a pinch of ascorbic acid
1 cup flax meal
2 cups water
1 tsp. vanilla
2 tsp. cinnamon
2 cups organic cane sugar

Mix all ingredients together. Do not over mix. Bake 45 minutes at 325°F.

[The following recipes are from Kentucky State University Pawpaw.][2]

Pawpaw Custard Pie

1 cup 2% milk
1 cup cream
3 eggs
¾ cup sugar
1 cup pureed pawpaw pulp

Mixing the ingredients as you add them, beat together the milk, cream, eggs, sugar, and pawpaw. Pour the custard into a pie shell and bake at 450°F for 15 minutes, then reduce the heat to 325°F and bake an additional 40 minutes or until a knife inserted near the center of the pie comes out clean.

Pawpaw Pie or Parfait

½ cup brown sugar
1 envelope unflavored gelatin (or agar agar)
½ tsp. salt
⅔ cup milk
3 eggs, separated
1 cup strained pawpaw pulp
¼ cup sugar

In a saucepan, mix together brown sugar, gelatin, and salt. Stir in milk and slightly beaten egg yolks. Heat and stir until mixture comes to a boil. Remove from fire and stir in pawpaw pulp.

Chill until it mounds slightly when spooned (20 to 30 minutes in refrigerator). Shortly before the mixture is sufficiently set, beat egg whites until they form soft peaks; then gradually add sugar, beating until stiff peaks form. Fold the partly set pawpaw mixture thoroughly into egg whites. Pour into a 9-inch graham cracker crust or into parfait glasses and chill until firm. "Then lock the door to keep the neighbors out."

Pawpaw Custard

1 cup pawpaw pulp
2 oz. grated coconut
1¼ cup half and half
1 tsp. vanilla
3 eggs, beaten
1 dash salt
2 oz. sugar (superfine preferred)
zest of orange (optional), grated

Mix pawpaw pulp with coconut. Layer on bottom of buttered ovenproof casserole dish. Heat half and half mixed with the vanilla until bubbles form. Beat eggs with salt and sugar and still beating, pour on the half and half very slowly so as not to curdle the eggs. Add the orange rind if using. Pour over fruit and place in a pan of hot water. Bake in a moderate oven (375°F) for 30 minutes or until custard is set. Turn out if possible when cool to show off the fruit layer.

Pawpaw Cream Pie

¾ cup sugar
⅓ cup flour or ¼ cup cornstarch
3 egg yolks, slightly beaten
1 cup milk
1 cup light cream
1 cup pureed pawpaw pulp
3 egg whites
3 Tbsp. sugar
pinch of salt
1 baked 9-inch pastry shell

Combine sugar and flour or cornstarch. Add the beaten egg yolks, milk, and cream. Mix well and add pawpaw pulp. Cook and stir constantly over low heat until thickened. Cool.

Make a meringue by beating the egg whites stiff with the sugar and a pinch of salt. Pour custard into a baked pastry shell and cover with meringue. Bake in a moderate oven (350°F) for 12 minutes or until meringue is browned. Serves 6 to 8.

Pawpaw Cookies with Black Walnuts

¾ cup pureed pawpaw pulp
1 cup all-purpose flour
½ tsp. baking powder
¼ cup butter
½ cup brown sugar
1 egg
½ cup black walnuts

Preheat the oven to 350°F and grease one large cookie sheet. Peel and seed fresh pawpaws and process in a food processor until fine. Sift together the flour and baking powder, and set aside. Cream the butter and sugar. Add the egg. Add the flour mixture and then add the pawpaw pulp. Chop half the nuts (reserve 16 pieces) and blend them in. Drop by teaspoonfuls onto the prepared cookie sheet and press a piece of black walnut onto the top of each cookie. Bake 12 minutes or until brown across the top. Makes about 16 cookies.

Pawpaw Ice Cream (Pure Amazingness)

1 quart cold milk
6 eggs
½ tsp. salt
1½ cup sugar
1 cup pureed pawpaw pulp, or more to taste
juice of 1 lemon
1 quart heavy cream
2 Tbsp. vanilla

Scald 3 cups of the milk in the top of a double boiler. Beat eggs well; add salt, sugar, and the remaining cup of milk. Stir egg mixture slowly into the hot milk and cook over a small amount of simmering hot water, stirring constantly, until mixture just coats a clean metal spoon. To prevent curdling, do not have the water boiling vigorously, and take care not to overcook. Stop cooking as soon as the custard coats the spoon and remove from heat at once. Cool pan of custard in another pan containing cold water, then chill thoroughly in refrigerator.

Combine pawpaw puree with the lemon juice and add to the chilled custard along with the cream and vanilla. Pour mixture into a chilled 1-gallon ice cream freezer canister and fit dasher into place.

Freeze and ripen according to directions accompanying ice cream freezer, or as follows: Fill the freezer tub around the canister with finely cracked ice and salt, using 1 part ice cream salt to 8 parts of ice, or about 1 quart of salt for a gallon-sized ice cream freezer. Fill the freezer half-full of ice before adding the first layer of salt, then alternate layers of ice and salt until the tub is filled. Freeze until the ice cream stiffens (about 20 minutes with an electric ice cream freezer). Then repack the freezer tub with ice, or remove the ice cream and place it in an ice cream mold, and let the ice cream ripen for several hours before serving.

To repack the freezer, remove the dasher, plug up the hole in the lid of the ice cream canister, drain out the salt water through the hole in the side of the ice cream freezer, and add fresh ice and salt to fill the freezer tub. Put cracked ice, but no salt, over the top of the canister, too. Cover the whole freezer with blankets or newspapers and let it stand in a cool place for several hours.

Ethnic Markets

This is perhaps the least explored, but perhaps one of the best potential avenues. Remember that *Asimina triloba* has many delicious and very similar tropical fruit tree relatives in such diverse places as India, most of South and Central America including Mexico, Thailand, Hawai'i, Malaysia, and beyond. With the US being a residence of so many ethnic groups, you have millions of immigrants that long for the fruits and flavors of their homelands. Fresh fruit is often more avidly sought out, utilized, and relished by people from tropical and third world countries than by the general population of the US. Tropical peoples are more accustomed to custardy, ultra-sweet fruit like pawpaw. I have given fresh pawpaw to numerous people from India and other tropical countries, and they all relished it and wanted more, and offered to buy all I had. Many were amazed that such a fruit grew here (Kentucky) and likened it to the *sapotes*, *sitapol,* and custard apples of their homelands. This is a good thing. If you live in a city with lots of ethnic people from tropical regions, and/or Latinos, by all means give them samples of pawpaws and set up shop. Talk to local ethnic stores, Mexican markets, and Indian grocery stores. Talk to local ethnic restaurants. They may seem skeptical or uninterested at first, but once they try it, they will usually love it and may become very enthusiastic. Make sure the pawpaws are high quality and very ripe. They will have to see and taste it to become interested.

This marketing strategy may take a few years to develop and will involve some education and outreach, but could develop a very steady and enthusiastic market for the pawpaw in ethnic areas. They may even want to come to your farm to pick it or buy a boxful. Try to get a few people very interested and enthusiastic, and they will often take care of the rest and become evangelists of the pawpaw for you. For high-quality fresh produce, some ethnic people will pay high prices.

11

Pawpaw Cultivars

In this chapter, I have attempted to compile every known pawpaw cultivar to be found as of 2020. There are so many obscure, unnamed, or numbered selections that tracking them down and compiling them is difficult and not necessarily very useful. The most obscure ones may be near impossible to acquire, and most of them are obscure for a reason.

Although the majority of these cultivars will not be found in the nursery trade, some could be available from collectors, many of whom are members of the NAPGA or NAFEX (North American Fruit Explorers). Also check with members of fruit growing forums online. Cultivars considered extinct or lost are listed in the following section.

Please note: Not all of these named available cultivars are desirable or exemplify good fruit characteristics. Like any fruit, there are poor-quality named cultivars in circulation. Many of these are even lauded in usually reliable nursery catalogs as excellent cultivars; yet they are actually anything but. Never rely on most nurseries to provide accurate and unbiased cultivar descriptions, especially for pawpaws. There has been a trend in the last few years, however, of many nurseries dropping older, usually inferior pawpaw cultivars and adopting the newer superior cultivars such as KSU's releases. This will make many of those old cultivars rare and vulnerable to extinction, something collectors should take note of. However, many older, poor-quality cultivars will eventually disappear, and that is all right as newer, far superior ones are on the horizon that can better meet the needs of the burgeoning pawpaw market and clientele.

While pawpaw breeding and selection is ongoing, the cultivars that are currently available are sufficient to create a viable market; successful merchants are benefitting from growing these select excellent pawpaw cultivars.

With that in mind, I have compiled and described nearly every known cultivar with as much available, generally unbiased, and detailed information from multiple sources as possible or useful. I have attempted to be fair and objective when describing flavor and texture (and have personally never tasted most of these cultivars), but when many reports identified bitterness or poor flavor, this is added to the description as generally reliable, as well as glowing reports of excellent quality from numerous reliable sources.

The information presented here is for educational and archival purposes only. Any choices you make based on these descriptions is at your own risk and may vary from region to region, and season to season. Tree health, age, and soil also affect the fruit quality from place to place. Remember, just like many fruits, pawpaws have their "vintage seasons" and also others when certain cultivars just don't perform like expected. One cultivar may perform excellently in one region and poorly somewhere else, which may be why descriptions vary among sources. However, the following information is generally accurate and has been confirmed by much research and confirmation from reliable sources when possible.

Key To Cultivar Descriptions

Cultivars that are listed with a star (*) are the most highly recommended for market growing (and usually make excellent backyard trees as well). These possess a combination of superior qualities: medium-large or bigger fruit size, good production of fruit, excellent flavor and pleasant texture. They often possess other noteworthy qualities, such as a thick durable skin or an early or late production season that makes cultivating them much more strategic than choosing cultivars based only on size or flavor. If planning to grow pawpaws commercially, you should focus on these recommended selections as your primary cultivars.

Codes: BEF = Blandy Experimental Farm collection (Boyce, VA); KSU = Kentucky State University in Frankfort, KY

Available: Currently available in the nursery trade

Ripening Season: To say "early ripening" or "late ripening" doesn't mean much without a reference. The ripening times below are generalizations based on typical ripening dates within the pawpaw belt (upper South and warmer Midwestern US), Zones 5b to 6b.

General pawpaw ripening dates described are as follows: "early ripening" (August 20–September 5), "mid-season ripening" (September 5–30) and "late season ripening" (October 1–31 and beyond). In addition, we will designate "very early ripening" as anything generally ripening before August 20. Some growers claim certain cultivars, such as 'Summer Delight', ripen as early as late July in central KY, and this is a good example of a "very early ripening" cultivar.

Ripening dates vary from year to year and depend on many factors: date of bloom due to spring temperatures, amount of heat days in a particular season, microclimate and sun exposure on the growing site, and cultivar blooming and ripening time. Some cultivars will bloom at the same general time as most others but ripen fruits weeks or even months earlier. These ripening dates are not guaranteed but are close approximations based on cultivar descriptions and grower information. Individual trees may vary substantially from descriptions of ripening time due to the combination of factors described above, but this can serve as a general guide under good growing conditions in a "typical season." It is safe to assume that in the pawpaw belt most fruit ripens between September 1 and October 1, with some exceptions. If you intend on commercially growing in the far northern ranges of pawpaw territory, such as Ontario, Canada, or upstate New York, then only choose cultivars known to ripen in your area or known to ripen early. Planting late ripening cultivars in these areas could result in a complete crop failure and waste, as frost will destroy the fruit still hanging on the trees. Plan carefully and accordingly.

Grading Scale: The grading scale is mostly geared toward market growing interests such as overall quality based on size, flavor, and texture, as well as reliability and production. Based on KSU taste tests, grower reports, personal experience, and many grower evaluations, I have graded each cultivar either Excellent (A), Good (B), Mediocre (C), or Poor (F), mostly in terms of *commercial viability*. You may disagree with the quality labeling. However, it is based on reports from many different sources, including expert pawpaw growers, nurserymen, and researchers, as well as my own observations, testing, and trials. I did not create the Grades based on my own personal opinions or guesses.

Also, note that for simplicity, size descriptions, if available, are generally as follows: jumbo (16+ ounces), large (12 to 16 ounces), medium (7 to 12 ounces), and small (3 to 6 ounces). In general, the larger the size and weight of the pawpaw fruit, the better it is for marketing and processing purposes. These tend to have much greater flesh-seed ratio (more fruit flesh and less percentage of seed weight) and are often simply better eating quality. Generally, any fruits in the small size range or smaller are not worth growing or trying to market. Where reliable quality information was not available or was vague, cultivars are marked "Quality unknown."

What about the pawpaws graded F? Why would someone name and propagate a terrible cultivar with poor quality? There are several possible reasons for this. First, it is speculated (but highly likely) that some cultivars, such as 'Wilson', were actually decent pawpaws, but through a nursery mistake or collector mishap, the actual cultivated tree got damaged and the scion died and left a seedling rootstock behind. That seedling rootstock (with inferior fruit quality) was then propagated and distributed as 'Wilson'. Again, an inferior tree could have also been misidentified with 'Wilson' somewhere along the line, and propagated as such. These things happen. Furthermore, in some areas, such as Florida or the Gulf area, pawpaws may be found growing wild occasionally, but often do not have very desirable fruit genetics, especially when compared to specimens found in the pawpaw's main range, including Kentucky, southern Ohio, Tennessee, Indiana. So, in those areas, any wild tree often seems like a desirable pawpaw and is propagated as such. Finally, sometimes novices and collectors will propagate cultivars that just aren't very good but have some sort of historical, cultural, or sentimental value (or are good seed producers for nursery rootstock). Whatever the case, any pawpaw graded F is to be avoided, and any graded C are highly suspect of being mostly worthless. When planning to grow pawpaw fruit commercially, strictly choose to plant only A grade and perhaps a few B grade cultivars for diversity or special interest.

NOTE: Cultivars marked with asterisks (*) are superior cultivars that have the best marketing and commercial potential thus far.

Pawpaw Cultivars and Descriptions

The following pawpaws are in existence as of 2019-2021 and can be found throughout nurseries, pawpaw collectors' orchards, farms, and university plantings. Some may be difficult if not nearly impossible to access, and many are not found for sale. Fortunately, most of the best ones worth planting are sold in nurseries. Not all are worth growing; they are listed only because they are currently in existence and may be obtainable.

'Allegheny': Vigorous cultivar that grows well and produces high-quality fruit with medium fleshiness; percent seed approximately 8% by weight. Fruit size approximately 4 ounces (125 grams) at KSU, unthinned. Flesh color is yellow, texture is medium firm and smooth, flavor is excellent, some say citrus-like. Fruit size can benefit from thinning the young fruitlets when fruit set is very high. Neal Peterson says fruit thinning is required on this cultivar to keep fruit size over 8 ounces. Needs some fruit to be removed to reach medium size and has higher seed content than other Peterson selections. Some say this is better as a backyard rather than commercial pawpaw due to smaller size and thinning requirement. That may be true, so planting large amounts of this cultivar should be expected to increase labor requirements in the orchard. B+. Available.

'Ark-21': Seedling pawpaw with reportedly good quality, grown by Louisiana State University. Not available in the nursery trade. Quality unknown.

'Atria': Heavy-bearing cultivar with large elongated fruit. Rare. Quality unknown.

'Baker's Acres #1': Chris Chmiel selection from Athens County, OH. Reportedly good. Quality unknown.

'Baker's Acres #2': Chris Chmiel selection from Athens County, OH. Reportedly good. Quality unknown.

'Belle': Originated from the orchard of Corwin Davis in Bellevue, MI. Fruit size is around 8 to 16 ounces (250 to 500 grams). Early variety. B.

*** 'Benny's Favorite' (VE-21):** 'Davis' × 'Prolific'. Mid-season ripening pawpaw (VE for "valley east," not "very early"), large, slight color break to yellow when ripe. High-quality fruit, average size is medium, 6 to 8 ounces. Very good rating by Jerry himself. Reports are very positive on the quality of the fruit. Jerry Lehman introduction. In our orchard, it has shown some susceptibility to black spot infection. A.

'Bertria': Heavy-bearing cultivar with large good-flavored fruit. Rare. Quality unknown.

'Broad': Chris Chmiel selection. Medium-large fruit with light-colored aromatic flesh. Originated in TN. Reportedly a poor producer with mediocre-poor fruit quality. Disappearing from the nursery trade. C.

'Catochin': Selected from the wild in VA. Supposedly high quality with orange flesh. Unavailable in the nursery trade. Quality unknown.

'Cawood': Firm texture, light-yellow flesh. Color break to yellow when ripe. Late-season ripening. Reportedly has very good quality and excellent flavor. Gets very high scores in KSU taste tests, similar in ratings to 'Susquehanna'. Medium-large size. Worth growing, but it's become rare in the nursery trade. It would be worth looking into this one. Older cultivar. A. Available.

'Cales Creek': Wild seedling from Summers County, WV. Quality unknown.

'Cantaloupe': Medium-large fruits around 10 ounces with melon flavor. According to Cliff England, "The tree has been observed with clusters of 3 to 4 pawpaw fruits that are over 300g each, falling and splitting into two clean pieces." Originated in Richmond, KY. Somewhat freestone. Woody Walker introduction. Quality unknown. Available.

'Car Wreck': Chris Chmiel selection. Apparently a car crashed into the mother tree at some point. Good producer. Quality unknown.

'Caspian': Newer release discovered in northwestern MO by Tyler Halvin and named after his youngest son. Very productive yielding 8- to 18-ounce fruits. Low seed weight. Taste is sweet and fruity, smooth textured with no bitter aftertaste. May be an excellent choice for short-season and northern ranges of pawpaw culture. Quality unknown. Available.

'Cherokee Ridge': Ron Powell early selection from the wild. Resists grafting. Small-medium size. He honestly graded it a D or F.

'Collins': Selected in GA. Quality unknown.

'Convis': Selected from Corwin Davis's orchard. Large fruit, averages about 10 to 15 ounces, yellow flesh; ripens first week of October in MI, probably ripens mid-September in KY. B.

'Cox's': No info. Unknown.

'Creek Ridge': Ron Powell selection. Small fruit but heavy production. Good source for rootstock seed as it produces lots and lots of seedy fruit. Not good for eating. C.

'Cullman Late': Don Cullman selection from Marysville, OH. Very late ripening. Medium-sized with good flavor. Hangs on the tree until October. Quality unknown.

'Danae's Creekside' ('Halvin's Danae's Creekside'): Good-quality fruit, but little information available. Newer selection. From Iowa, so it's likely very hardy and earlier ripening. Quality unknown. Available.

'Danika': From Yellow Springs, OH. Selected for large fruit. Quality unknown.

'Davis': Selected from the wild in MI by Corwin Davis in 1959. Introduced in 1961 from Bellevue, MI. Medium-sized fruit, up to 5 inches long and average about 4 ounces; green skin; dark-yellow-orange flesh; large seed; ripens first week of October in MI; keeps well in cold storage. Similar to, yet smaller than, 'Sunflower'. Hardy to -25 F. C. Available.

'Dr. Chill': A selected seedling from 'Overleese' seed. Medium-large fruit. Good quality, but seedy. Nice name! Named after a friend of the grower, originated in PA. Unknown.

'Early Surprise': Oriana Kruszewski selection from her farm in Skokie, IL. Apparently has exceptional cold hardiness and ripens early enough to not get damaged by autumn frosts in northern regions. Said to be hardy to Zone 4b. Not currently available in the nursery trade. Quality unknown.

'Estil' ('Estill'): Selected in Frankfort, KY. Large fruit with smooth texture. Medium-large size. Reportedly good quality. C. Available.

'Ford Amend': Selected in the Pacific Northwest and grown since 1950. Green-yellow skin and flavorful orange flesh. Probably mediocre, but may be well-suited to the Pacific Northwest and other difficult pawpaw growing regions with cool summers. Once thought extinct, was discovered growing in OR. Might have BSD (Blue Stem Disease) resistance that annihilates most pawpaws planted in the Pacific Northwest. C.

'Fort Hunter': From Pennsylvania, reportedly has good quality. Quality unknown.

'Franklin Valley Gold': Selected in Pike County, OH. Good producing with good flavor. Medium-size fruit. B+.

'Gainesville #1' and '#2': Grafted from trees found growing in Gainesville, FL. Fruit appears small and likely is very mediocre in fruit quality. May perform better in Gulf states and Deep South and may also lack the cold hardiness of more northerly cultivars. Quality unknown. Available.

'Gene': From Adams County, OH. Medium-large fruit with good flavor. Resilient. C.

'Glaser': Selected in Evansville, IN. Medium-size fruit. Noted in 1974 by CRFG. C.

'Georgia': An Italian self-fertile variety that ripens in Torino, Italy, in late August and early September. It is described as a "very aromatic, delicious fruit with yellow-orange pulp that contains 10 seeds." This cultivar is patented in Europe, and therefore is extremely difficult to obtain in the US at this time. Fruit size is small. Late ripening in October. Quality unknown.

'Ghent 2 Blom': From the botanical gardens in Ghent, Belgium. Seeds originally came from the Morton Arboretum in Iroquois, IL. Medium green color with a buttery fruit pulp. Very aromatic, sweet, delicious taste. Matures in the second week of October in Ghent, Belgium. May not be available outside of Europe. Quality unknown.

'Golden Moon': Discovered growing in Berea, KY, by the author. Named in honor of the Bengali Bhakta saint Chaitanya Mahaprabhu who was known also as the 'Golden Moon'. Fruits ripen early season. Fruits are medium-large, around 8 to 16 ounces on average. The fruits are oblong and plump with medium to low seed count. Thin-skinned; flesh is thick, marshmallow-like, and very sweet, with excellent texture and flavor, similar to 'Sunflower'. Heavy producer and is hardy to adverse conditions. Somewhat susceptible to *Phyllosticta*, which blemishes the fruit; however, it handles the fungus well, without cracking. Ripe fruit turns a very noticeable bright golden yellow at ripening (color break), that helps in knowing when to harvest. Introduced by Peaceful Heritage Nursery in 2018. B+. Available.

*** 'Greenriver Belle':** Discovered near the Green River in Hart County by Carol Friedman in 1998. Neal Peterson and Ron Powell both agree this is one of their favorite cultivars and say it has wonderful flavor. Said to be productive, high-quality, and have no bitterness or aftertaste (some claim a cinnamon aftertaste). Medium size. Little to no *Phyllosticta*. Few seeds. Good producer. Firmer texture than most. Good shipping potential. Short shelf life. Good reports about quality. A. Available.

'Hale #5': Selected by George Hale of Columbus, OH, in IN. Best Fruit Winner at the 2004 Ohio Pawpaw Festival. Medium-large fruit with good flavor. There are also 'Hale #1-4'. Quality unknown.

*** 'Halvin' ('Halvin's Side Winder'):** Wild seedling discovered in southwestern Iowa, near Bedford, by Tyler and Dana Halvin. Reportedly was discovered as a 40-foot wild understory tree. Said to have excellent flavor and medium-size fruit. Mild pineapple-like flavor and aroma with good texture and no aftertaste are reported. 'Halvin' is a hardy and resilient variety that performs well under adverse soil and climatic conditions. May be suited to trial or breeding for northern regions or those outside the typical pawpaw growing range due to cold-hardiness and early ripening. 'Halvin' is well-liked by those who have tried it. A. Available.

'Hardy Wonder': Oriana Kruszewski selection from her farm in Skokie, IL. Apparently has exceptional cold hardiness. Said to be hardy to Zone 4b. Not currently available in the nursery trade. Quality unknown.

'Hayes': Medium-sized fruit with good aroma and taste. May not be available in the US. Unknown quality.

'Honey Dew': Medium-large fruits around 9 to 10 ounces. Has melon flavor and thick texture. Originated in Richmond, KY. Somewhat freestone. Woody Walker introduction. Quality unknown. Available.

'Hoberg': Medium size, with good aroma and flavor. Selected in Europe by Hans Brinkmann–Landwehr. May not be available in the US. Quality unknown.

'Hoot Owl': Variegated cultivar from Athens, OH. Has ornamental landscape tree potential. Slow-growing. Not for fruit production.

'Hopi Ridge': Ron Powell selection from Adams County, OH. Very heavy producer. Decent flavor. C.

'Horn's White' ('Al Horn Whiteflesh'): Seedling selected by the late Al Horn of Indiana, who was an avid pawpaw grower and marketer. Creamy textured, yellow-skinned fruit with lovely stark white flesh. Flavor is sweet and coconut-like. White-fleshed pawpaws are uncommon but often have exceptional quality. Said to have very good quality and large to possibly jumbo fruits. In our trials, it has been difficult to establish. Rare. A. Available.

'Ithaca': Originated in Italy, selected by Domenico Montanari. Medium-size fruit, said to be good quality and heavy bearing. May not be available in the US. Unknown quality.

*** 'IXL' (may go by 'ISL' or '1 XL'):** 'Overleese' × 'Davis' hybrid selected in Eaton Rapids, MI. Large fruit, around 12 ounces. Yellow flesh; ripens second week of October in MI, probably end of September in KY.

Late ripening. Well-shaped and handsome tree. Good flavor. Gets high ratings in taste tests, and growers seem to like this one a lot. A. Available.

'Jack Irvine #1': Mediocre fruit quality. C.

'Jenny's Gold' ('KSU G6-120'): KSU selection. Quality unknown.

'Jeremy's Gold': KSU selection. Mediocre taste test results. Quality unknown.

'Jerry's Big Girl' ('250-39'): Large to jumbo fruit up to 24 ounces! Ohio Pawpaw Festival Winner for Biggest Fruit. Very good flavor reported. 'Sunflower' × 'Sam Norris-15'. Large-jumbo 10 to 16+ ounce fruit with banana-like flavor, very creamy, soft texture. Shows yellow skin-coloration when ripe. Thin skins make it quite delicate. Mid-late season in KY. Reports of over five pounds of fruit from one peduncle! Medium seed count. Jerry Lehman selection, well-liked by him. A. Available.

* **'Jerry's Delight' ('250-30'):** Open pollinated seedling of 'KY-8-2'. Won the Biggest Pawpaw contest in 2013, 680 grams and placed third in the Best Pawpaw contest 2014. Great flavor and very sweet. Jerry Lehman introduction. Reportedly licensed outside the US. A. Available.

'Jonathan': Yellow flesh. Slight aftertaste. Small-medium fruit. Unknown quality.

'Kirsten' (may go by 'Kirstin'?): Hybrid of 'Taytwo' × 'Overleese'. Selected by Tom Mansell of Aliquippa, PA. Mediocre-low quality. C.

'Kentucky Champion': Discovered in the early 2000s in Madison County, KY, by Woody Walker. The original tree was registered with the KY Dept. of Forestry as the largest pawpaw tree in the state. Very early-ripening, long, banana-shaped fruits average about 8 ounces. Said to be hardy and resilient, and Woody thinks it may perform well in cooler, shorter-season areas. The seed-to-pulp ratio is about 8.3%. Introduced by England's Nursery. It is reportedly shy bearing which makes its usefulness for commercial production questionable. Needs more evaluation. Now trademarked overseas. Quality unknown. Available.

'Kings Gold #2': Great name, apparently mediocre fruit quality. May have originated in Ohio. C.

* **'KSU Atwood' ('KSU 8-2'):** KSU's first named pawpaw selection, released in 2009. Originated from seed sent to KSU from MD. A very noteworthy and highly desirable cultivar. High yields, delicious fruits and *very strong disease resistance* to *Phyllosticta* all define this superior pawpaw. Fruit has greenish-blue skin and bright, golden-yellow flesh with low seed count. Fruit size is medium (averaging 209 grams). KSU reports 150 fruit (about one bushel) per tree at KSU. Smooth clean skin, attractive bluish color; excellent, very refreshing bright tropical fruit flavor and nice creamy custard texture. Medium seediness with small seeds. Precocious, starts blooming at 4 to 5 feet tall, long bloom season. Good commercial potential. This is one of the author's favorite pawpaw cultivars. Good commercial pawpaw cultivar, but the fruits are not huge. The superior appearance and productiveness of this pawpaw are its major standouts. Average weight 7.5 ounces, 10.3% seed, and 24 Brix. A+. Available.

'KSU Benson' ('KSU 7-5'): Second recent release by KSU in autumn 2016. This superior variety features *very good disease resistance* to *Phyllosticta* and simply has a wonderful, extra-sweet rich flavor and very thick buttery texture that is outstanding. Flavor is somewhat persimmon-pumpkin-cherry with a nice, thick custardy texture and low seed count. Medium-size fruits, some may be large. Fruit shape is often almost spherical. Flesh color is light orange and attractive. Good commercial potential. Jeremy Lowe at KSU thinks 'KSU Benson' may do better early in its productive life, and fruit size and consistency may go down as the tree reaches 15 to 20 years old. In our experience, it's not the easiest or fastest cultivar to establish in the orchard or to graft, but the amazing quality and productiveness makes it worth it. Fruit size varies from 4 to 12 ounces (sometimes more), 9% seed, and 16.7 Brix. A. Available.

'KSU Chappell' ('KSU 4-1'): 'KSU Chappell' produces good size fruit, around 8 to 12 ounces with some 1-pounders in there. Thought to be a seedling of 'Susquehanna'. Flavor is outstanding. The skin is thick and durable, but not as much so as 'Susquehanna'. The flesh is thick and buttery like 'Susquehanna', but the flavor is less intensely sweet and rich. Tropical overtones, honey and caramel blended together in this complex yet smooth, refreshing wonderful flavor. Very low seed count similar to 'Susquehanna'. Has some *Phyllosticta* resistance. Some growers have concerns about the fruit cracking. It has great potential and makes very high-quality fruit but needs more testing in different areas. Shows extremely vigorous and strong hybrid growth, reaching 4 to 5 feet after a couple of seasons from grafting. Average weight 8 ounces, 4.9% seed, and an incredible 27 Brix, making it very sweet indeed! A+. Available.

'Lady D': From MI. Medium fruits average about 10 ounces (300 grams). Said to have good flavor. Quality unknown.

'Louisiana Native': Small-medium fruit. Yellow flesh with good light-caramel flavor. Originated in Louisiana, so likely a good candidate for growing in the Gulf States and Deep South. Fast growing, vigorous. Suitable for the Gulf States but better cultivars are available for elsewhere. Quality unknown.

'Lehman's Chiffon': 'Prolific' × 'Sunflower'. A seedling of one of Jerry Lehman's selections. Said to have very good sweet flavor and creamy texture with heavy production. Quality unknown. Available.

*** 'Lehman's Delight' ('275-48'):** Late-season producer. Large-jumbo fruits of good quality. 'Prolific' × 'Sam Norris-15' hybrid. Seems to have a lot of potential. Very large fruit with good flavor, soft texture. Won the Ohio Pawpaw Festival's Biggest Pawpaw contest in 2011 and 2012. Won second place in the Best Pawpaw contest in 2013 and 2014, beating out 14 other entries each year. Mid- to late-season harvest; excellent sweet flavor. Jerry Lehman introduction. A. Available.

*** 'Lynn's Favorite':** Selected from seed from Corwin Davis's orchard. Yellow-fleshed, medium-large fruit; ripens second week of October in MI. Reportedly has very good-quality fruit. Won Best Tasting Fruit at Ohio Pawpaw Festival. Soft texture flesh with thin skin. Very productive. A. Available.

'Mammoth': 'Overleese' × 'Sunflower' hybrid. Large-jumbo fruit of excellent quality. Unknown quality.

'Mango': *Very vigorous cultivar*; grows more rapidly than most known cultivars. However, in the author's trials, *it appears to take the same amount of time to fruit as other cultivars* (3 to 5 years), but will often be much taller and wider than other cultivars of the same age at that point. Fairly good flavor, potentially better suited to hotter and longer-season areas of the South due to its Tifton, GA, origination. The texture of 'Mango' is usually watery or even slimy and unpleasant; however, some people think it is excellent. The texture becomes softer and juicy as the fruit over-ripens. The looser juicier texture is said to make 'Mango' excel as a processing variety. Does not keep well. Has a fairly tropical flavor profile. Medium-size fruit. Reports of bitter aftertaste. 'Mango' may be more useful for breeding vigorous strains and for interstems/rootstock than for fruit production. Personally, I find it fairly low-quality eating, and the texture is simply awful (it oozes out into your hand). However, the fruits do attain good size, and the vigor of the tree is impressive. Not highly recommended for planting. B-. Available.

✱ 'Maria's Joy' ('166-13'): Jerry Lehman cultivar. It is a 'Davis' × 'Prolific' cross, formerly well-known as '166-13'. This cultivar has a solid reputation among pawpaw people because of large fruit size, a mild and delicious tapioca/mango flavor, and good production. Reliable yields of 8- to 14-ounce, medium-large fruits. Won the Biggest Pawpaw Contest at the Ohio Pawpaw Festival in 2012. Well-suited to commercial production. Strong grower. Excellent flavor. A. Available.

'Mark Twain': Older cultivar, extremely rare. Quality unknown.

'Marshmallow': Very sweet flavor and firm, marshmallow-like texture. Medium fruits around 7 to 9 ounces. Originated in Richmond, KY. Somewhat freestone. Woody Walker introduction. Quality unknown. Available.

'Mary Foos Johnson': Selected from the wild in Kansas by Milo Gibson. Seedling donated to North Willamette Experimental Station, Aurora, OR, by Mary Foos Johnson. May have 'Sunflower' parentage. Medium-large fruit averages about 8 ounces; yellow skin; butter-color flesh; few seeds; ripens first week of October in MI. Mediocre quality, low production. C. Available.

'M. Gordon': Wild seedling from Cranbury, NJ. Quality unknown.

'Middletown': Selected from the wild in Middletown, OH, by Ernest J. Downing in 1915. Ripens in mid-September in Kentucky. Fruit size small; averaging 75 grams (2. 5 ounces) and reports of 75 fruit per tree at KSU. Mediocre quality. Small fruit size makes it nearly worthless except to collectors. Neal Peterson thinks this is a case where the original scion died and a low-quality rootstock took over. C.

'Mitchell': Selected from the wild in Jefferson Co., IL, by Joseph W. Hickman in 1979. Fruit has a slight color break to yellow at ripening, golden flesh; light, fluffy texture and good flavor, few seeds. Fruit size small-medium, averaging 115 grams (4 ounces) and 60 fruit per tree at KSU. Good flavor, but overall mediocre due to smaller fruit size. Older cultivar that has been surpassed by better, newer ones. C.

'Mohican': Quality unknown.

*** 'Overleese':** Selected in 1950, discovered on the Overleeses's property by W.B. Ward in Rushville, IN. Produces large fruits of good mild flavor and good texture with few seeds. An older variety and still considered one of the best pawpaws for both commercial and backyard growers. Said to produce excellent offspring ('SAA Overleese' and 'Shenandoah' are both 'Overleese' offspring). Reports say it would make a good shipping pawpaw. Highly recommended. A. Available.

'Neal': Good producer of sweet fruit. Quality unknown.

*** 'NC-1':** Hybrid seedling of 'Davis' × 'Overleese'; selected by R. Douglas Campbell, Ontario, Canada, in 1976. Fruit is large, has few seeds; yellow flesh; thin green skin; early ripening, September 15 in Ontario and early September ripening in KY. Fruit size large; averaging 180 g/fruit and 45 fruit per tree at KSU. Considered not only one of the best varieties in general but also *has excellent potential in northern regions*, and for developing other early-ripening, cold-hardy, and resilient northern cultivars, due to its Ontario origination. Fruit texture and flavor are excellent, smooth, buttery, and very sweet. Ron Powell says this cultivar is consistently robust and healthy wherever he finds it growing, with large, disease-free dark-green leaves. Excellent quality, productivity, hardiness, great flavor, and nice thick texture make 'NC-1' a top cultivar. Highly recommended. A. Available.

*** 'Nyomi's Delicious':** Original tree discovered in Berea, KY, tended to by Nyomi, an elderly local herbalist woman who described her backyard tree as a "magical cancer-curing fruit." Nurseryman Cliff England went to Nyomi's house to investigate and found the mysterious "magical" fruit to be an excellent local pawpaw. 'Nyomi's Delicious' produces exceptional quality fruit. Color break to light yellow at ripening. Very sweet, pleasant mild flavor with no unpleasant aftertaste that will be acceptable to people who find other pawpaws too strong-tasting. Soft, creamy, thick, dense, excellent texture with a light-yellow pulp with low seed weight. Not goopy or slimy. This is a reliable heavy producer of rounded, nicely shaped, 8- to 16-ounce fruits that are 4 to 8 inches long. Hangs in cluster of 4's and 5's but seems to often produce single hanging fruit. Average skin thickness, but otherwise has good commercial appeal due to mild, very sweet flavor and large size. *Leaves and fruit skins are clean and free of* Phyllosticta *when annually observed by the author in Berea, KY,* and in other locations shows resistance. The tree takes well to grafting and grows vigorously. This is one of the author's favorite cultivars and deserves greater recognition and should be widely planted. 'Nyomi's' has it all: excellent quality flavor and texture, reliable and heavy production, good fruit size, single peduncle bearing habit, color break at ripening, and disease resistance. May have some issues with surviving through its first winter in the ground, so plant bigger specimens. Good work finding this one, Cliff! A+. Available.

'October Moon' ('275-17'): Late-season pawpaw, ripening in late September or early October. Ohio Pawpaw Festival winner. Heavy bearer, large fruit with good flavor. Good for season extension. Jerry Lehman introduction. A.

'Pepper 1': Selected by Chris Chmiel. Reportedly selected from the largest known pawpaw tree in OH. Mediocre quality. Quality unknown.

'Pepper 2': Selected by Chris Chmiel. Quality unknown.

'Pickle': Small fruit, with pickle shape. Mediocre quality. Rare. C.

* **'Pennsylvania Golden' Series:** The 'Pennsylvania Golden' pawpaws are very suited to growing in the far northern reaches of the pawpaw range, such as Pennsylvania, Ontario, and northern NY. The medium to large fruits have an attractive appearance and beautiful flesh.

They would also be useful to stretch the season out by providing an early season cultivar, but growers in other regions have better choices available. There is some confusion about the different strains of 'Pennsylvania Golden' as most nurseries do not differentiate between the numbers and simply use the name 'Pennsylvania Golden'. Figuring out exactly which one they're selling is near impossible. Available.

'Pennsylvania Golden' ('PA-Golden #1'): Seedling originated from George Slate collection by John Gordon, Amherst, NY. Very productive cultivar, but fruits are often small-medium, occasionally large. Fruit has yellow skin when ripening, golden flesh; matures late August in KY and mid-September in NY. Fruit size medium; averaging 110 grams (3.8 ounces) and 120 fruit per tree at KSU. Good quality and may perform better than most in the pawpaw's northernmost range, such as NY. Author and pawpaw grower Lee Reich says the 'PA Golden' series are among his favorites. B.

'Pennsylvania Golden #2' ('PA-Golden 2'): Selected as seedling from seed originating from George Slate collection by John Gordon, Amherst, NY. Fruit: yellow skin, golden flesh; matures mid-September in NY. Good quality and may perform better than most in the pawpaw's northernmost range, such as NY. B.

'Pennsylvania Golden #3' ('PA-Golden 3'): Selected as seedling from seed originating from George Slate collection by John Gordon, Amherst, NY. Fruit: yellow skin, golden flesh; matures mid-September in NY. Very good quality and may perform better than most in the pawpaw's northernmost range, such as NY. Some consider this the best 'PA Golden' cultivar. B+.

'Pennsylvania Golden #4' ('PA-Golden 3'): Selected as seedling from seed originating from George Slate collection by John Gordon, Amherst, NY. Fruit: yellow skin, golden flesh; matures mid-September in NY. Good quality and may perform better than most in the pawpaw's northernmost range, such as New York State. B.

* **'Potomac':** Considered a very good cultivar with excellent fruit. Some reports of fruit split as an issue, which leads to rotting and yield loss and/or *Phyllosticta* infection. Large-jumbo fruit typically 12+

ounces. Has a delicious sweet flavor and pleasant thick texture. Has *Phyllosticta* susceptibility, as do many other cultivars. Extremely fleshy with great thick texture. Percent seed ~ 4% by weight. Medium productivity. Strong apical dominance; i.e., the tree grows very upright, is less spreading than most. Neal Peterson says, "Dr. Pomper (KSU) insisted I release this." A. Available.

'Prairie King #1': Introduced by pawpaw winemaker Ken Neighbors of IN. Won Best Tasting Award at the 2007 Ohio Festival. Quality unknown.

'Prairie King #2': Introduced by pawpaw winemaker Ken Neighbors of IN. Won 2nd Place Best Tasting Award at the 2007 Ohio Festival. Quality unknown.

*** 'Prima 1216' ('Montanari 1216'):** Fascinating cultivar originated out of a 1983 Faenza, Italy, planting of 1,000 seeds of US-grown Corwin Davis pawpaw seed (some sources say 'Sunflower' cultivar), imported and planted on the farm of Domenico Montanari. Out of his trial planting, 'Prima 1216' was selected as producing the biggest, best-quality fruit. Fruit commonly attains large-jumbo size with excellent quality; texture is very soft, silky and buttery and not goopy. The flavor is good, very mild, sweet, and pleasant. One grower reports an "off-flavor," and I also detected a very slight yeasty undertone; yet I found the overall flavor and texture very good. Has jet-black large slender seeds and low seed weight, about 5%. Very fleshy. Skin is thick and rubbery and exceptionally durable. A good candidate for trial plantings for commercial production. *Appears resistant* to *Phyllosticta* infections. Performs well in parts of Italy and other parts of Europe, including Croatia. Performs better than average in KY. Vigorous and very fast grower, very precocious, and comes into bearing early. In our trial planting, it produced 10 pounds of excellent fruit in its fourth year. Overall, 'Prima 1216' is a highly recommended cultivar with many excellent qualities. A. Available.

'Prolific': Selected by Corwin Davis, Bellevue, MI, in mid-1980s. Medium, 7- to 8-ounce fruit, yellow flesh; late ripening, first week of October in MI. Fruit size medium at KSU. High-yielding and precocious cultivar, but quality is generally considered mediocre, although some think it is excellent. Decent flavor, may have coconut overtones and thick, dense texture. Reports of bitter aftertaste many find unpleasant. Has been useful for breeding purposes, used extensively by Jerry Lehman. B. Available.

'Quaker Delight': Originally from the Wilmington College Arboretum in OH, original tree is now gone. Won the Best Fruit Competition at the Ohio Pawpaw Festival in 2003. Has a mild flavor; light-yellow flesh and creamy texture. Medium fruit. Early-mid-season production (early September in OH). One of Ron Powell's favorites. B+.

'Rana': Won Best Fruit in 2004 at Ohio Pawpaw Festival. Quality unknown.

'Rappahannock': High-quality fruit with pleasant aftertaste. Durable skin good for handling and marketing. Vigorous and fast-growing cultivar. Large beautiful, well-shaped, and symmetrical fruits, whether born singly or in clusters. Fruits are about 6% seed by weight. The flavor is superior as well, very sweet and refreshing. Receives mixed reviews around hardiness and productivity. It's somewhat easier to harvest with horizontally held leaves revealing the fruit and a slight color change at ripening to yellow. Many growers consider this the lowest quality of Peterson's selections, although opinion varies, and some consider it a high-quality cultivar. Has a tendency toward making poor crotch angles or poor branching structure leading to branches breaking. Color break to yellow when ripe. Fruit tends to be small, 3 to 5 ounces in KY and the Midwest; yet in MD, VA, and NC, it tends toward being medium-large. *Susceptible to* Phyllosticta *infection*. Even though it's a Peterson selection, it reportedly does not perform well in many situations. Reports from North Carolina say it performs well there. Fails to perform in the Midwest. For all these reasons, it's not a highly recommended cultivar but is worth trialing in some areas, notably North Carolina. B. Available.

'Rebecca's Gold': Possibly selected from Corwin Davis seed by J.M. Riley in 1974, apparently originated in Northern California, others say Indiana. Medium-large fruit, kidney-shaped; yellow flesh.

Fruit size medium at KSU. Reports of this cultivar doing well in Northern California are reported. Fruits are mild and sweet, yellow flesh is somewhat juicy to watery and not the best. Thin skin, medium-low seed weight. Sometimes this cultivar is reported to produce large fruits (1 pound or more), but average tends to be 3 to 6 ounces. Ripens mid-season. Very sweet flavor, but due to smaller fruit size, thin skin, and overly soft, juicy texture, it is not highly recommended. May be good for processing. B. Available.

'Regulus': Heavy-bearing cultivar with medium-large fruit from 8 to 16 ounces. Low seed count with very good flavor. Very rare. The nursery that introduced this cultivar no longer sells it. Quality unknown.

'Rigel': Large fruit, 8 to 16 ounces with low-seed weight and good flavor. Very rare. The nursery that introduced this cultivar no longer sells it. Quality unknown.

'Ruby Keenan': Medium-sized fruit with "excellent flavor" according to KSU. Other reports say it is mediocre in quality. Size appears medium. May be extinct. B.

'Sam Norris' Cultivars: Sam Norris was a Kentucky plantsman and pawpaw breeder. He attempted to create tetraploids by using a plant hormone called colchicine. These (possible yet unconfirmed) tetraploids selections exhibited unusually large flowers and leaves and have been used in breeding better pawpaws. He named several cultivars including 'Sam Norris #15', used by Jerry Lehman in breeding. Neal Peterson says diploids and tetraploids rarely crossbreed, and so he doubts these are true tetraploids. Quality unknown.

✻ **'SAA Overleese':** Selected from 'Overleese' seed by John Gordon, Amherst, NY, in 1982. Large fruit. Rounded shape; green skin; excellent flavor yellow flesh; few (but large) seeds; matures in mid-October in NY, probably late September in KY. A. Available.

✻ **'SAB Overleese':** Selected from 'Overleese' seed by John Gordon, Amherst, NY. Large fruit; rounded shape; green skin; yellow flesh; few seeds; matures in mid-October in NY. *May have some* Phyllosticta *resistance*. Late ripening. A. Available.

✻ **'SAC Overleese':** Selected from 'Overleese' seed by John Gordon,

Amherst, NY. Large fruit; rounded shape; green skin; yellow flesh; few seeds; matures in mid-October in NY. A. Available.

* **'SAA Zimmerman':** Selected as seedling from seed originating from G.A. Zimmerman collection by John Gordon, Amherst, NY, in 1982. Large fruit; yellow skin and flesh; few seeds. Color break to yellow when ripe. Excellent, sweet flavor. Early ripening, thin skin, soft textured fruit. Author and grower Lee Reich says this is one of his favorites. A. Available.

* **'SAB Zimmerman':** Selected as seedling from seed originating from G.A. Zimmerman collection by John Gordon, Amherst, NY, in 1982. Large fruit; yellow skin and flesh; few seeds. Color break to yellow when ripe. Excellent, sweet flavor. Early ripening, thin skin, soft textured fruit. *May have some* Phyllosticta *resistance.* A.

* **'SAC Zimmerman':** Selected as seedling from seed originating from G.A. Zimmerman collection by John Gordon, Amherst, NY, in 1982. Large fruit; yellow skin and flesh; few seeds. Color break to yellow when ripe. Excellent, sweet flavor. Early ripening, thin skin, soft textured fruit. A.

'Sarah G': No info. Unknown.

'Seneca Ridge': Ron Powell selection. Reliable and productive, medium fruit. C.

'Shawnee Trail': Dick Glaser selection from Wilmington, OH. Best Fruit Winner in the 2002 Ohio Pawpaw Festival. Yellow flesh, good mild flavor. Average 5 oz fruit. B.

* **'Shenandoah':** Has good market acceptance due to its delicious flavor and smooth texture. Many like the mild tropical/banana flavor, but it's usually not described as having excellent flavor (Neal Peterson reports his Latino customers found it similar to cherimoya, favorably). Good production, reliable. Medium-large size. Superior variety all around, highly recommended for commercial production. Texture is custardy; flavor is smooth with just the right balance of fragrance, sweet fruity flavor, and agreeable aftertaste. This tree originated as a seedling of 'Overleese', and according to Neal Peterson, "it must be said, is superior to its parent." Large fruit with few seeds (6% by

weight). Fruit is often born in single-fruited clusters. Good yields. 'Shenandoah' responds well to pruning. A+. Available.

'Sibley': Supposedly a sibling/sport of 'Sun-Glo', 'Sibley' produces large fruit (up to 12 to 14+ ounces/400 grams), with golden yellow pulp of good quality and a color break to yellow when ripe. Available in Europe and possibly the US. C.

'Simina': A newer introduction from Italian and Romanian growers. Fruits around 7 ounces, whitish-yellow flesh. Has a creamy texture and medium seediness. Not currently available in the US. Quality unknown.

'State Fair' ('H3-120'): Unreleased KSU selection. Tends to make single-fruit clusters. Medium-large fruit. Good quality. A.

'Sue': Yellow-fleshed, 4- to 6-ounce fruit. Flavorful fruit and a productive tree. Originated in southern Indiana, discovered by Don Munich. Quality unknown.

'Summer Delight': A Cliff England introduction, apparently ripens earlier than any other known pawpaw cultivar. Very sweet, refreshing, pleasing tropical flavor and smooth texture. Very early-ripening cultivar could stretch the pawpaw season considerably and create marketing advantages. *Very early ripening also may make this a very desirable cultivar to trial in northern regions.* Unfortunately, in our orchard, it has been difficult to establish as well as propagate and seems possibly prone to black spot. B. Available.

'Sun-glo' ('Sun Glow'): Yellow skin, yellow flesh, large fruit. Ripens first week of October in MI, probably mid-September in KY. Quality unknown.

*** 'Sunflower':** Discovered wild in Chanute, KA, by Milo Gibson in the 1970s. A strong, vigorous grower with showy large flowers. Hardy and resilient with reliable fruit production. The medium-jumbo fruits average around 8 to 12 ounces, but can be over a pound each. Flavor is complex, caramel-vanilla-nutty with a mild fruity aroma. Texture is thick, puffy, marshmallow-like and creamy; excellent. Medium seed weight with large flat seeds, more seedy than most other highly recommended varieties. Has a slightly bitter flavor near the skin and seeds that some find disagreeable. Has stood the test of time. High yielding. Displays a degree of self-fertility. Some claim 'Sunflower' seeds make excellent rootstock. 'Sunflower' would be a very good commercial cultivar due to impressive size, high yields, and excellent texture and flavor. Has been used in KSU's breeding programs. 'Sunflower' is one of the author's favorite cultivars, and in a good year, is absolutely incredible. A. Available.

'Super Mario': Also known as 'NN10-35'. Quality unknown.

*** 'Susquehanna':** This unique pawpaw shows the potential that breeding and selection can attain. Thick durable skin protects the fruit when harvesting, shipping, and handling. Resembles a commercial mango. Good yields. 'Susquehanna' has a firm, luscious, buttery, avocado-like texture and a rich honey-like caramel sweetness. Too intensely sweet and strong for some, the favorite of others. Keeps exceedingly well, even at room temperature. Picked just barely underripe, they slowly ripen at room temperature for a week or more in fine condition, and can keep in refrigeration for a month. Very low seed weight that can be very easily removed when a fruit is halved (3% seed by fruit weight!), with some nonfertile, aborted seeds often present. 'Susquehanna' is perhaps the best-suited commercial cultivar in overall quality due to its thick and durable skin, especially firm avocado-like flesh, long shelf-life, and outstanding flavor. Has great marketing and shipping potential. Leaves are especially attractive. Reports of dieback or lack of vigor seem to be its main drawbacks. Perhaps crossbreeding it with more vigorous but also excellent cultivars might create a similar, even superior, cultivar (this is indeed what KSU has done). One of the author's favorite cultivars. *Prone to* Phyllosticta *in humid, wet summers.*

A drawback to 'Susquehanna' is that many growers find that it does not seem to have a high survival rate the first year after transplanting, and it is not especially vigorous, often growing slowly. A great pawpaw and first-class fruit, but be cautious when planting lots of them; probably best to plant larger specimens and be prepared to replace some of them. A+. Available.

'Sweet Alice': Old variety, circa 1945. Medium-large fruits with orange-yellow flesh. Good flavor and quality, no aftertaste. *May have some* Phyllosticta *resistance.* Hardy in Zone 5. Originated in WV by Homer Jacobs. May be well-suited to the South, does well in AL. Unknown quality. Available.

'Sweet Potato': Chance seedling from the Blandy Experimental Farm at the University of Virginia. The medium-large fruit is yellow-fleshed and has a pleasant sweet flavor. Ripens August through September. Does well as far south as in northern FL. Quality unknown. Available.

'Sweet Virginia': Originated in MI. Small-medium, sometimes large fruits up to 12 ounces, considered possibly self-fertile. Good flavor. Quality unknown.

*** 'Tallahatchie':** Released in Autumn 2018, 'Tallahatchie' is the latest Peterson release as of this writing. We tasted it in 2019 and thought it had a complex, exotic flavor. 'Tallahatchie' is distinctive: pleasant aroma, sweet mellow flavor with floral notes and a very smooth texture. Very low seed weight. 'Tallahatchie' tends to bear large clusters that require thinning to achieve best size. Yields are medium-high. Ripens mid-season to late. Fruit averages 9 ounces with very few seeds (seed to fruit ratio is 5%). KSU staff have insisted it "must always be included" in making pawpaw ice cream due to the excellent flavor. Fruits are plump and attractive. If it's anything like Peterson's other selections, it is very high quality. Neal Peterson says: "This is fine in a backyard setting, but for orchards is problematic. Pawpaw fruits do not ripen simultaneously within a cluster; so repeated harvest of a cluster must be made over a period of days. Too often a fruit falls from a cluster seconds after a fruit is picked." A. Available.

'Taylor': Selected from the wild in Eaton Rapids, MI, by Corwin Davis in 1968. Fruit has green skin, yellow flesh; ripens in September in KY and first week of October in MI. Fruit size medium, averaging 3.8 ounces and 70 fruit per tree at KSU. C. Available.

'Taytwo' (also 'Taytoo'): Selected from the wild in Eaton Rapids, MI, by Corwin Davis in 1968. Fruit has light-green skin and yellow flesh; ripens in September in Kentucky and first week of October in MI.

Fruit size medium-small, averaging 4.2 ounces and 75 fruit per tree at KSU. Originated from the same small patch of trees as 'Taylor'. Outstanding tropical sweet, fruity flavor profile with caramel and cherimoya overtones. Medium-high seediness is very noticeable. Vigorous tree. Fruit have an excellent appearance. People will like the flavor on this one, but the smaller fruit size may limit its commercial usefulness. B. Available.

'Tollgate': Yellow-fleshed, ripens first week of October in MI, probably mid-September in KY. Corwin Davis selection. Late ripening. Good flavor, productive when mature. Small-medium fruit around 6 to 8 ounces. Old variety that has been surpassed. C.

'Tropical Treat': Selected from the wild in KY by Woody Walker in 2010. Medium-sized fruit, up to 6 inches long, weighing 6 to 10 ounces (180 to 300 grams). Yellow-green skin at ripening, yellow flesh, large seed, ripens mid-August in KY, keeps well in cold storage. Has a thick skin. Flavor is reportedly excellent, with mango and pineapple overtones. Introduced by England's Nursery in 2015. Quality Unknown. Available.

'UVM #1': 'Sunflower' seedling. A very cold-hardy seedling selection that reportedly fruits in Burlington, VT. I've never heard of a pawpaw fruiting this far north, so may be worth checking out if you want to grow pawpaws in VT or other northern areas. Small-medium size fruit. Unknown.

'Vickey Russell': Won 2nd place for Best Pawpaw at Ohio Pawpaw Festival in 2004. Wild selection. Quality unknown.

'Victoria': No info. Unknown.

* **'Wabash':** Some consider this the best-performing Peterson cultivar. Good tree health and vigor, medium-large fruit averaging 6.5 to 12 ounces, and heavy production (65 fruit per tree reported at KSU). May have some self-pollination potential. Sweet, rich tropical-caramel flavor overtones. Fruit size will benefit from thinning. Very fleshy; percent seed ~ 6% by weight. Texture medium firm, creamy, smooth. Flesh color yellow to orangish. The overall fruit quality and quantity on this variety is excellent. A favorite of those who have tasted it at the

KSU orchard. Neal Peterson adds, "Dr. Pomper also insisted I release this one." A+. Available.

'Walter's': No info. Quality unknown.

'Wells' ('Well's Delight'): Selected from the wild in Salem, IN, by David Wells in 1990. Fruit has a green skin, orange flesh. Ripens mid to late-September in Kentucky. Fruit size medium, averaging 105 g/fruit (3 ounces) and 65 fruit per tree at KSU. Fruit may get up to 10 to 12 ounces. Won the 1990 KSU Pawpaw contest. Mediocre quality and small size make it unappealing except to collectors. Some say 'Wells' has "large, tasty fruit." Some 'Wells' trees available may possibly be mixed cultivars going under the same name, and some are apparently better than others. Likely this cultivar has been lost due to scion wood collection error, as its current form is not the prize-winning tree of former years. C. Available.

'Wilson': Selected in 1985 from the wild on Black Mountain, Harlan Co., KY, by John V. Creech. Fruit has a color change to yellow at ripening and has golden flesh. Ripens in September in Kentucky. Fruit size small; averaging 2.75-ounce fruit and 130 fruit per tree at KSU. Mediocre quality and small size make it nearly worthless except to collectors. Reportedly terrible quality; humorously, Dr. Kirk Pomper once said when eating 'Wilson' fruit "you may wish for death"! However, one of its redeeming qualities is that, in one of the Corvallis, OR, trial plantings, 'Wilson' had a 100% survival rate while most of the other pawpaw trees were decimated by BSD (Blue Stem Disease). This could make 'Wilson' very useful in breeding BSD resistant cultivars. It remains mostly useless for fruit production. F.

'Whip Longsderf': Selection from PA. Quality unknown.

'White Owl': Selection from White Owl Nursery in IL. Quality unknown.

'Yuri's Russian': From Russia. Quality unknown.

'Zimmerman': Selected in NY from G.A. Zimmerman seed by George Slate. Good-mediocre quality. Medium fruit at KSU. Difficult to obtain, rare. Quality unknown. Available.

Numbered Pawpaw Selections Worth Mentioning

'G6-120' (Nicknamed 'Jeremy's Gold'): KSU selection. As Jeremy Lowe describes it: "This selection has a strong caramel flavor even when under ripe. It also has a thicker skin with russeting, which might help with shipping. Fruit size and seed weight aren't very impressive. Average weight 4.7 oz, 9.3% seed, and 22.1 Brix." Gets mixed ratings in KSU taste tests.

'G9-108': KSU selection. Receives very high ratings for flavor. Quality unknown.

'G9-109': KSU selection. 147 g average weight, 11.2% seed, and 26.6 Brix. Quality unknown.

'G9-111': KSU selection. Average weight 89 g, 13.7% seed, and 19.1 Brix. Quality unknown.

'Haz-1': KSU selection. Quality unknown.

'H-3-120' ('State Fair'): KSU selection, not released. Grafts take well, and scion grows well. Medium productivity. Average weight 214 g, 6.3% seed, and 25 Brix. Quality good.

'KSU HI 1-4': Being tested at KSU as a potential new cultivar. Shows a low seed weight, good sweet flavor, yellow flesh. Likely a 'Susquehanna' × 'Sunflower' hybrid. Large fruit and medium seed content. Grafts take well and scion displays good growth. It's fairly susceptible to *Phyllosticta* (about like 'Sunflower'). Average weight 180 g, 9.8% seed, and 27.1 Brix. I think this cultivar is very high quality.

✻ 'KSU HI-7-1': Excellent seedling tree at KSU. KSU is considering releasing this one. Fruits are long and dense, banana shaped, and around 8 to 12 ounces; some may get bigger. The flesh is thick, very light blonde, almost white, and has this outstanding cherimoya × banana flavor that is stellar. Researchers at KSU thinks the fruits are higher in seed weight than they would prefer their releases to be. Otherwise the quality is outstanding. Average weight is 200 g, 11.6% seed, and 25 Brix. B+.

'KSU 27': Early- to mid-season production. Quality unknown.

'KSU 2-11': Very late-season production. Quality unknown.

'KSU G4-25' ('Pina Colada'): Unofficial name for an unreleased KSU selection. Has a mild, very delicious pineapple-coconut flavor. Hardiness and survivability are currently being tested by KSU. Not likely to be released due to survivability issues and difficulty in grafting. Average weight 216 g, 7.5% seed, and 22.3 Brix. I first witnessed this fruit at KSU in 2019, and it was extremely impressive, with massive gorgeous fruits of the finest quality. The flavor was very pleasing, mild and sweet, with excellent thick texture. Excellent size and fruit quality, but propagation difficulty is the major issue holding it back.

'K 8-2': Released as 'KSU Atwood'. See listing for 'KSU Atwood'.

'VE-5': Very early-season ripening, apparently, but mid-September ripening in northern IN. Jerry Lehman introduction. Quality unknown.

'VE-9': 'Davis' × 'Prolific'. Large fruit, late season, high quality. Jerry Lehman introduction. Quality unknown.

'166-20': 'Overleese' × 'Prolific', with advantages in sweetness and yellow fruit color for easy ripe detection and marketing. About 8-ounce fruit. Good, sweet flavor. Jerry Lehman introduction. Quality unknown.

* **'166-66':** 'Davis' × 'Prolific', with advantages in sweetness and yellow fruit color for easy ripe detection and marketing. Average 6 to 8 ounces. Good flavor. Jerry Lehman introduction.

'275-25': Jerry Lehman selection. My personal tasting notes read: "Tropical, delicious flavor. Light-yellow flesh. Banana/wild pawpaw flavor, creamy texture."

'275-50': 'Prolific' × 'Sam Norris-15'. Similar to '275-48'. Excellent pollinator for '275-48'. Jerry Lehman introduction. My tasting notes read: "Very creamy, smooth, nice flavor. Texture a little watery. Tropical, slight caramel flavor; refreshing, light flavor with low seed count. Dark-yellow flesh." I rated it 7/10.

* **'275-56':** 'Prolific' × 'Sam Norris-15'. Average-large fruits 6 to 8 ounces. Heavy crops with banana-like, wild pawpaw flavor. Soft, creamy texture leaning toward watery. Has won awards in the Ohio Pawpaw Festival Contests. Jerry Lehman introduction. Delicate skin needs careful handling. A.

'275-56N': Jerry Lehman introduction. May be superior to "275-56S". Quality unknown.
'275-56S': Jerry Lehman introduction. Quality unknown.
'275-60': 'Prolific' × 'Sam Norris-15'. Late bearing. Average to large, good quality. Jerry Lehman introduction. Quality unknown.
✱ **'275-69':** Won second place in the Best Pawpaw Contest at the 4th International Pawpaw Conference, Frankfort, KY, 2016. Medium fruits are buttery, smooth flesh, very sweet. Has light-yellow skin color break when ripe. Jerry Lehman introduction. A.
'400-27': Jerry Lehman selection. Quality unknown.
'400-50': A hybrid of *A. triloba* × *A. incarna* bred by Lester Davis, Columbus, GA. Open pollinated. Large flowers, small fruit, *inedible.*
'400-275': Jerry Lehman selection. My tasting notes read: "Banana/melon flavor. Creamy with vanilla undertones. Large seed sacks. Nice golden-yellow color. Top choice. 8/10."

Missing/Extinct Pawpaw Cultivars

The following pawpaw cultivars were available at one time but are now considered "missing" and likely extinct. I simply wouldn't put any relevance whatsoever on trying to search out and locate any of these old cultivars. It would be near impossible to track down the few that might still exist; most if not all are long dead and gone, and overall most of them likely had/have mediocre fruit quality compared to the newer releases. This list is included simply for scholarly purposes and historical information. As for planting, just focus on the best currently available cultivars and plant seeds from these to start new, even better cultivars.
'Arkansas Beauty': Originated in AR as a wild tree.
'Betty Wirt': Large fruit up to 1 pound, but averages 6 ounces. Wild selection from Wirt County, WV.
'Buckman': Late-ripening fruit with white flesh. Mild flavor. Selected by Benjamin Buckman of IL c. 1910.
'Cale's Creek': No information available.
'Cheatwood': Selected in the wild of Gallia, OH, by J. Cheatwood.
'Cheely': Wild selection from Iuka, IL, by J. Cheely.

'Cox's Favorite': Selected from wild.
'Dr. Potter': Selected in Julet, IN, by B.S. Potter. Small fruit; mild flavor; ships fairly well; late maturity; rich yellow flesh.
'Duck': No information available.
'Early Best': Selected in wild in IN by W.C. Stout.
'Early Cluster': Selected from wild.
'Early Gold': Selected from wild.
'Endicott': Selected in Villa Ridge, IL, in wild by G. Endicott.
'Fairchild': Early maturing. Selected by famous plant breeder David Fairchild from seed of the cultivar 'Ketter'.
'Fairchild #2': Seedling of 'Fairchild' grown at the Blandy Experimental Station.
'G-2': Selected from G.A. Zimmerman seed by John W. Mckay in College Park, MA, in 1942.
'Gable': Late to very late maturity, selected in wild in PA by J. Gable.
'Hann': Selected in the wild in AR.
'Hengst': Selected from wild.
'Holt': No information available.
'Holtwood': Selected from wild by W. Hoopes.
'Hope's August': Early maturing. Selected from wild in Paint, OH, by A. Hope.
'Hope's September': Selected from wild in Paint, OH, by A. Hope.
'Jumbo': Late to very late maturity.
'Kercheval': No information available.
'Ketter': Matures evenly; skin comparatively thick and tough; does not discolor markedly; flesh medium yellow; mild but rich flavor, neither insipid nor cloying; large yellow fruit; early maturity. Said to have been very excellent quality. Selected in Ironton, OH, by Mrs. F. Ketter. Prize in a Journal of Heredity pawpaw contest in 1916.
'Kurle': Small-medium fruit; yellow flesh and skin. Seedling of 'Davis' grown by R. Kurle in MI in 1982.
'Lawvere': No information available.
'Little Rosie': Small fruit, reported to be an excellent pollinator. Selected by R. Glaser in Evansville, IN.

'Long John': Selected by J. Mckay from 'G-2' seed in College Park, MD, in 1948.

'M-1': Selected by J. Mckay from 'G-2' seed in College Park, MD, in 1948.

'Martin': Large fruit (Zimmerman said small size); flesh yellow and of superior quality (Zimmerman said skin tough); withstands cold well. Selected in Springfield, OH, by J. Mckay from 'G-2' seed.

'Mason-WLW': Selected in wild by Ernest J. Downing in Mason, OH, in 1938.

'Mudge': Selected from wild.

'Osbourne': Late to very late maturity, selected from wild.

'Oswald': Selected from wild by E. Oswald in Hagerstown, MD.

'Probst Early': Selected from wild.

'Rees': Flesh pale yellow and of good flavor; seeds exceptionally small; not a large fruit size. Selected from wild by W. Rees in Hagerstown, MD.

'Roach': This appetizing-sounding pawpaw was selected by J.C. Roach in Dekalb, MO.

'Schriber': Selected from wild.

'Scott': Selected from wild in WV by C.S. Scott.

'Shannondale': Late to very late maturity.

'Silver Creek': Medium-size fruit. Selected from the wild by K. Schubert.

'Talbot': Fruits 10 ounces; flesh yellow; overall quality average. Chance seedling of Corwin Davis seed selected by John Talbot in Linton, IN.

'Taylor': An old variety also named 'Taylor' but not the more recent available cultivar of the same name.

'Uncle Tom': Probably the first named cultivar on record (assuming the Native Americans never named any!). Ripens mid-September in IN; fruit sets singly and in pairs. Selected from wild by J.A. Little around 1896 in Cartersburg, IN.

'Van Der Bogart': Very similar to 'PA-Golden'; matures mid-September in Ithaca, NY. Selected by Francis Van Der Bogart around 1970 from seed originating from the G.A. Zimmerman collection.

'Vena': Possibly the same as 'Talbot'. Possibly from Linton, IN.

35 Extinct Pawpaw Cultivars from the Early 20th Century

The year represents the date of origin of the cultivar.
Source: Neal Peterson, petersonpawpaws.com

'Uncle Tom'	1896
'Cheely'	1900
'Hahn'	1900
'Early Best'	1900
'Arkansas Beauty'	1900
'Scott'	1900
'Endicott'	1900
'Hope's August'	1900
'Hope's September'	1900
'Cox Favorite'	1900
'Early Cluster'	1900
'Propst Early'	1900
'Ketter'	1900
'Cheatwood'	1900
'Martin'	1900
'Rees'	1900
'Oswald'	1900
'Potter'	1900

'Roach'	1900
'Fairchild'	1925
'Long John'	1930
'Taylor'	1930
'Tiedke'	1930
'Shannondale'	1930
'Osbourne'	1930
'Buckman'	1930
'Schriber'	1930
'Holtwood'	1938
'Hengst'	1938
'Gable'	1940
'Jumbo'	1940
'Betty Wirt'	1960
'Mudge'	1960
'Lawvere'	1960
'Kercheval'	1960

12

Troubleshooting, Cost Analysis, and Calendar

See chapter 9 for full information about disorders, diseases, and insect pests of pawpaw. Below is a quick-reference guide for common disorders.

Troubleshooting Guide for Pawpaws

My newly planted trees only grew a few inches the first season. This is normal and should not worry you. However, it is likely caused by some amount of root disturbance (transplant shock, this is common) and/or nutrient deficiency.

Solution: Be very, very careful when transplanting pawpaw trees; this should only be done from March–June, with April/early May being the ideal time. They have delicate roots that are easily disturbed or damaged. Shocked trees will grow very little the first year. Also, pawpaws cannot compete with grass or weeds and need thick, wide mulch application or otherwise intense and complete grass/weed suppression. If in their second season in the ground, they continue to grow extremely slowly, then you either have them planted in an unfavorable site, such as thick, wet clay soil or the soil fertility is lacking. Add diluted fish emulsion (about 1 to 2 cups per 5 gallons of water, give each tree 1 gallon of this solution every 2 weeks from April 1 to July 1) and/or add horse or cow manure to the base of each tree or a general balanced organic fertilizer formulated for trees. Remember, however, in the first season of planting, pawpaw trees often grow only a few inches. The next season, they should begin growing much more rapidly (1 to 2 feet per year) if they get the above care just described.

The leaves of my trees appear bleached, whitish and are wilting. Sunburn and/or dehydration.
Solution: Make sure all seedling trees under 30 inches and any grafted trees under 18 inches receive sun protection. Give trees adequate irrigation. If the trees are in a greenhouse, the issue may be whiteflies. Apply pyrethrum spray, focusing the spray under the leaves, and/or release ladybugs and increase ventilation and air movement. Make sure trees are not stressed or drying out in pots. If the trees are in a greenhouse and the leaves appear whitish or bleached looking, then the tree might be infested with thrips, whiteflies, and/or spider mites.

My pawpaw trees leaves are yellowing and defoliating earlier than they should (or earlier than the other pawpaw trees in the orchard). This tree is distressed, either through drought, lack of nutrition, root problems, or weakening due to black spot infection or damage. If it is simply dehydrated, water heavily and make sure next year it gets proper irrigation. If it is any of the other factors, keep an eye on it next season, and if it appears distressed once again, remove the tree. Some cultivars are more prone to black spot than others, and it can weaken trees quite bit.

All my trees are wilting and dying. Could be Asian ambrosia beetle damage, sunburn, dehydration, BSD, or a waterlogged soil or drastically off-balance pH.
Solution: This can only be avoided beforehand by proper site selection and setup. Before planting, check pH and adjust it to be about 5.5 to 7. Make sure the site is not waterlogged and does not have an extremely high water table before planting your orchard there. If dehydration or drought is ascertained to be the problem, provide regular irrigation and keep the trees well mulched. Make sure trees are cleared of weeds and any grass is kept at least 2 feet away from each tree. If you live in the Pacific Northwest (PNW), your trees may be afflicted with blue stem disease (BSD). Cut a branch off and split it. If the inner wood has a sickly bluish-black stained look to it, BSD is

likely the issue. There is no cure, and the orchard is likely doomed. 'Wilson' and 'Ford Ammend' cultivars may have some resistance to BSD. You could remove all existing pawpaws and replant with those cultivars. Best practice would be to replant on another site, or if you are in the PNW, consider growing something more adapted to your region. If upon close inspection, rows of tiny pin-like holes are all over the trunk, then this is most likely Asian ambrosia beetle damage. The trees are probably doomed and need to be pulled up and removed.

My mature trees have lots of little pencil-sized holes in the trunk, in horizontal rows, almost like a drill was used. Sapsuckers are at work. These are little woodpecker-like birds that feed on tree sap. Mostly harmless, however if they are feeding heavily, this may weaken trees over time.
Solution: Thick coats of 50% diluted indoor latex paint on trunks keeps them away. Burlap bags wrapped around and tied to trunks will help.

The bottom of the tree trunk is ripped and girdled looking. Do you hire string-trimmers to weed whack your orchard? Sounds like string trimmer or other mechanical damage. The trees are done for and need replaced. Even a half-girdled pawpaw tree is likely now toast.
Solution: Make sure all your trees have chicken wire or hardware cloth around the base. Voles sometimes damage fruit trees, and such cages will keep them out. Trees can recover from minor damage, but if they are bark-girdled completely or even mostly, they will die. You can attempt *bridge-grafting* them, but it must be done immediately. Look in grafting books and online for information on this. Long-term prevention is needed in the form of wire cages around trees and/or no string trimmers anywhere near the trees.

My fruit has black blotches and cracks, and/or is splitting open on the tree. This damage is from *Phyllosticta*, a common pawpaw disease.
Solution: Plant *Phyllosticta*-resistant cultivars such as 'KSU Atwood'.

Experiment with using compost tea sprays or other organic sprays, as listed in chapter 9. Preventative antifungal sprays (either organic or conventional) will help (although none are registered for pawpaw, so be careful). Once infection is visible, not much can be done. Remove and destroy any cracked fruits. Keep the orchard clean, free of rotting fruits; perhaps prune the trees to allow for good air movement and make sure the trees are getting good, drying sun and the orchard is free of weeds.

My trees are losing leaves early in the season, and/or making small leaves and producing few fruits. Sounds like the trees are either starved for nutrition or are old and in decline.
Solution: Give lots of nutrition in the form of buckets of compost, manure, bonemeal, seabird guano or rock phosphate, worm castings, etc. Utilize kelp extract applied as a foliar spray. If the trees are close to or more than 20 years old, they may be in decline and a new orchard should be planted. The old trees can be used for harvesting scion wood and providing pollination for a few years, but should be removed eventually. Grafted pawpaws seem to go into decline after about 15 to 20 years of production, in most evaluation trials.

The young trees have some broken branches and/or the bark is stripped or torn off. This is likely deer rub from a buck with antlers.
Solution: Start hunting or hire deer hunters, put up 8-foot fencing, or at least have chicken wire around the tree trunks. Deer seem to be averse to latex paint on the tree trunks. Any pawpaws planted in deer territory absolutely must have protection on the trunk, either a plastic tree tube or chicken wire.

The bark on the south side of the trunk is splitting. This is winter sun damage. The bark warms up on sunny winter days, expands, and then refreezes at night, shrinks, and splits in the process.
Solution: White *indoor* latex paint mixed in equal parts with water and painted all over the trunk prevents this from occurring. This must be

done annually or at least biannually. Unfortunately, latex paint is not usually allowed in Organic Certification. Kaolin clay products may be used instead, check with your local certifying agent.

Leaves look mottled, discolored, yellow, or veiny. This is likely a nutrient deficiency, sometimes of magnesium.
Solution: Add kelp extract as a foliar feed, apply a few tablespoons of Epsom salts to the drip line, and also apply manures and compost. Symptoms should not reappear next season, but current affected leaves may not fully recover.

Leaves look wrinkled, deformed, or misshapen. This occurs sometimes and the exact reasons are not yet understood. If it happens every year, remove that tree and burn it. It may be a virus.

I have a bunch of wild trees but zero fruit. Most likely this is a clonal patch of one or maybe two pawpaw trees that have suckered underground and created many additional trunks, appearing to be many trees, sometimes hundreds. If it has no other genetically different tree to pollinate it, then it will make no fruit, even though there appears to be many trees.
Solution: You could plant new pawpaws around it, or graft many of the suckers to different cultivars. Whatever the case, a genetically different pawpaw must be present to pollinate it, or no fruit will be produced. You could also try hand-pollination, bringing in pollen from a genetically different tree.

I have mature size pawpaw trees planted of various cultivars but little to no fruit. Pawpaws generally start to flower and fruit when about 6 feet tall, although flowering can begin on smaller grafted trees. If nutrition is poor or the trees are unhealthy, they may not flower.
Solution: Add phosphorous and trace minerals. Or, late spring frosts could be killing off the blooms every year. If you live in an area not known to support pawpaws, the climate could be unsuitable. Give

nutrition and time and try to ascertain if frosts are killing off the blooms in March to May. If they are, consider putting in orchard heaters. Also look into very hardy early-ripening cultivars in marginal pawpaw growing regions.

Many of my pawpaw flowers are falling off. This may be the handiwork of pawpaw peduncle borer *Talponia plummeriana*. See chapter 9 for more information and controls. Some years these insects can reduce the crop by up to 50% to 75%. Usually damage is minimal in a healthy orchard. Also, did a late spring frost event occur that could have burned off the blossoms? Severe storms and winds could dislodge blooms as well.

Cost Analysis and Timetable

How much does starting a pawpaw orchard cost? That depends on many factors, including the size and the cost of the trees you're planting. We're going to assume you have a suitable site already available. Let's dive in.

As of 2020, high-quality potted, 2-year-old grafted pawpaw trees sell for an average price of about $30 each. Usually 2-year-old trees are all that is available. A few nurseries offer older, larger trees, but it's not recommended that you plant anything older than 3 years old and more than 3 to 4 feet tall. Remember, pawpaws are sensitive to transplanting, and larger trees just do not establish or perform as well. You have to let go of any sense of rushing this process along for whatever reason. Plant small trees and be happy, and patient. The fruit will come in time, faster than you think.

Tree numbers per acre vary by the distance between your rows. KSU says you can fit 295 trees per acre,[1] assuming you follow their recommendations for planting density of 8 feet between trees and 18 feet between rows. They also recommend, at around 18 to 20 years, removing every other tree in the row.[2] This allows sun penetration into the base and lower portions of the trees and prevents competition for light, water, and nutrients. Trees can be planted in closer rows if you are mowing with equipment smaller than a tractor. However, rows should not be closer than 15 to 16 feet for adequate sun penetration, access, and airflow.

If you are planting 295 trees per acre at $30 each (including taxes or shipping if ordering online), then that equals $8,850 per acre. It's likely, however, that you could get a wholesale price from the nursery for that many trees, somewhere in the $15 to $25 range. That would be highly advisable to try to do. In addition, you will need tree protection, black plastic mulch, an irrigation setup, tools, and fertilizer.

A total of 885 feet of 2-foot-high chicken wire (approximately 3 feet for each tree tube) is needed for 295 trees. Alternatively, you could put up an 8-foot deer fence for a similar price, or much cheaper if you cut your own black locust trees or cedars for posts. Some people use plastic fencing, which is very cheap (yet not very durable): **$1,376**

Woven black (or white) plastic row cover (a 4-feet-by-4-feet square plus 6 staples per tree): **$508**

Irrigation (3,000 feet of ¾ inch orchard tubing): **$680**

590 drip emitters (2 per tree) **$195**

Digital timer for irrigation setup: **$200**

Organic fertilizer (premade blend for establishment year, 1 pound per tree): approximately **$200**

Quality hand pruners: **$50**

Establishment cost total per acre: **$6,258** to (wholesale tree price) **$10,683** (retail tree price)

Obviously, the largest expense by far is the cost of the trees themselves (approximately 295 per acre): $4,425 (wholesale price of $15 per tree) to $8,850 (retail price of $30 per tree). You could also plant high-quality seedling trees that are usually less expensive, or start the orchard by grafting your own trees on free or cheap seedling rootstock, thus saving a large sum of capital. Other costs reflect general online prices for equipment including drip tubing and plastic mulch cloth. You may have access to local outlets or produce supply houses that have cheaper prices. You could reduce the supply costs by 25% if you shopped in person at a local produce supply store. This cost analysis does not include other necessary items like mowers, string trimmers, gasoline, equipment maintenance, water bills for irrigation, and land. You will have to figure out those details based on your particular situation and resources. If you are short on land, consider planting on a (very good) friend's or relative's land, or a long-term lease.

In summary, with a spare acre, good rainfall, and lots of patience, you could establish an acre of non-irrigated, unprotected pawpaws that you grew from seed for pennies. Whether it would thrive long-term depends on the care you give it as well as deer populations, summer weather, and your karma. However, under no circumstances should you try to save money by planting inferior trees or trees that are stressed, damaged, or grown by poor quality nurseries.

Production Costs

About production costs, Kentucky State University says:

> Production costs (2018) for a mature pawpaw planting are estimated at $1,650 per acre, with harvesting and marketing costs at $6,200 per acre. These are estimated costs for a representative Kentucky planting and count a hired labor charge of $12.50 for pruning and harvesting. Total variable costs per acre, including interest on operating capital, come to approximately $8,400. Presuming gross returns of $9,600 per acre, returns to land, capital and management are approximately $2,200 per acre. These returns could be substantially higher than the assumed $1.75 per pound wholesale price.[3]

These figures seem to assume you're not doing any of the labor yourself. Let's say you're doing half the labor, and we can cut that cost in half to $3,100 per year. Then, assuming you can get $2 per pound for the fruit, gross returns would be $11,000 per acre. After expenses, your net returns would be $6,250 per acre. But, before you think this is your get-rich-quick scheme, remember, pawpaw farming is somewhat risky right now due to low customer demand, regional difficulties in production, and fluctuating prices for fruit based on demand and the market. It's always better and safer to assume, on average, you'll net half of what the numbers say, which would be $3,125 per acre at $2 per pound. Much better returns than corn and soy per acre, but probably you should not quit your day job. However, pawpaws can be a good addition to an existing operation, a good niche crop, and a valuable asset to your farming operation, if you're willing to dedicate the time to caring for the trees and carefully harvest and market the fruit.

Nonetheless, if you simply want to add 20 trees to your farm or a small backyard pawpaw orchard, you can certainly do that for less than $1,000. A backyard setup should only cost between $100 and $200, depending on the trees you get, how many and how much they cost, and how much you put into place to maintain them (mulch, cages, irrigation, etc.). However, for the commercial market grower, please do not put in your orchard without any weed control (organic or synthetic mulch) and tree protection. Also crucial is an effective irrigation system or means to water the trees. Remember, one severe drought can kill your whole young orchard. One string trimmer cleaning an orchard of unprotected trees can unintentionally girdle and destroy them all. Don't risk it.

Pawpaw Orchard Calendar

This is a basic quick-reference guide for the general annual maintenance of a pawpaw orchard. Exact dates and times differ within the pawpaw's main growing range. Growers in northern regions may need to add one month to the spring recommendations, and recede one month on the autumn recommendations.

Winter: November–February. Make sure trees older than 3 to 4 years have their trunks painted by November to avoid winter bark damage. Put deer and vole protection in place. Remove broken or distressed branches or any dead or heavily declining trees. Prune any trees in late winter to reduce height or remove problematic branches. Make sure to paint any newly exposed areas of the trunk made when pruning the trees, to avoid winter sun damage. Review the marketing strategies you used that year. What can you do better? What worked great and what didn't and why? Take cuttings of dormant scion wood from late February to late March, store in plastic bags in the fridge, moistened slightly and labeled carefully. Make inquiries with tree nurseries now, in November–December.

Spring: March–May. Hand-pollinate any trees in flower or bring in fly attractants to increase pollination. Begin fertilizing trees in late April, renew mulch, and weed. Spray for fungal diseases if needed. Graft any tree seedlings after all danger of frost has passed, in late April or early May, as pawpaw trees begin budding. Make sure your irrigation system, if being used, is in perfect working order and make repairs as needed before summer heat arrives. Sometimes animals chew holes in pipes (or mowers slice them), so check for leaks. Remove or mow down any emerging root suckers. Remove any shoots growing below grafted portions of the trees. Keep orchard mowed. Plant new pawpaw trees at this time. Replace UV protection tubes by April; make sure any newly planted trees needing UV protection have protection by April or early May before they fully leaf out.

Summer: June–August. Any pawpaw tree planting should be completed by June at the very latest. Extra water and care will be needed for later planted trees. Thin any fruit clusters to achieve single-peduncle fruits if needed. Spray for fungal diseases as needed. Remove any dead or dying trees and burn. Keep irrigating and fertilizing as needed. Stop fertilizing by August. Watch for insect infestations such as webworms and ambrosia beetles. Keep the orchard mowed and weeded. Remove or mow down any emerging root suckers. Remove any shoots growing below grafted portions of the trees. Keep vines off trees.

Very early season and early season fruits will start ripening around August. Create hype by early August if you are marketing pawpaws, making sure your marketing plans are solid and all your connections are aware pawpaws will be ripe soon. Check that processing equipment is in good working order and clean. Any marketing supplies should be on hand (stickers, bags, boxes, labels, etc.).

Autumn: September–October. Main pawpaw harvest season! Make sure you are handpicking fruits carefully when ripe for premium fruit sales. Keep up on harvesting daily. Remove rotted, split or unusable fruits from the orchard; don't let them simply rot on the ground or branches. Keep an eye on varmints in the orchard; trap or have dogs on the guard. Enjoy the harvest. Save seeds for starting seedlings next year for planting out or for rootstocks. Make sure the seeds are thoroughly cleaned, disinfected, and stored moist under refrigeration (34° to 38°F.) By mid-October, pawpaws are generally all but gone. Some very late cultivars (such as 'Prima 1216') may not ripen till October. Don't harvest these early! Late-ripening cultivars planted inappropriately in northern areas will likely not ripen before frost. In their main growing region, pawpaws completely finish weeks before the first frost. *Remove any UV protection plastic tree tubes before frost.* Save them and reuse next year, placing them back onto any trees shorter than 30 inches (for seedlings) or 18 inches (for grafted trees) by April. Paint your trees with a 50-50 water/white interior latex paint blend to prevent bark splitting and winter injury.

13

Conclusion

Pawpaws have strong potential to become a popular, healthy addition to the temperate world's diet and a lucrative, relatively easy-to-grow crop for small, organic, and other farmers. Pawpaws may prove to be more popular in Europe and Asia than in North America. In my opinion, they need more refining through breeding and selection in order to meet a few important needs, namely, lowering concentrations of acetogenins that likely lead to nausea in some people, as well as improving fruit consistency, durability, disease and crack resistance, size and overall quality. We're not that far off from these goals, but work still remains.

After many years of working with pawpaws and observing them in dozens of natural/wild as well as heavily managed and cultivated scenarios, I am convinced the fragile trees are most adapted to grow as a natural, multi-trunked, suckering colony and not as pruned, single trunk, standalone trees, which reduces their lifespan drastically and makes them susceptible to many ailments. These include winter damage to the trunk, ambrosia beetles, and extreme weather events. However, standalone trees are the easiest way to manage an orchard and what most growers will continue to do for some time.

The future of pawpaws will undoubtedly focus on utilizing tissue cultured/cuttings-propagated trees, which would have the great advantage of being capable of producing suckers true to type instead of suckers from genetically inferior rootstock (as with grafted trees). Horticultural scientists in Europe[4] are already perfecting the exceedingly challenging process of tissue culturing pawpaws successfully.

This would allow the trees to produce multiple trunks and shoots, thus making the lifespan of the trees drastically longer and allowing them to recover from even the severest of damage to the aboveground growth from insects, climatic events, accidental damage, etc. Even if all the top growth was lost, the trees could recover rapidly and begin bearing fruit in a few years, with no need to replant. We should not fight the pawpaw tree's natural inclinations and adaptive strategies, but work with them in concert.

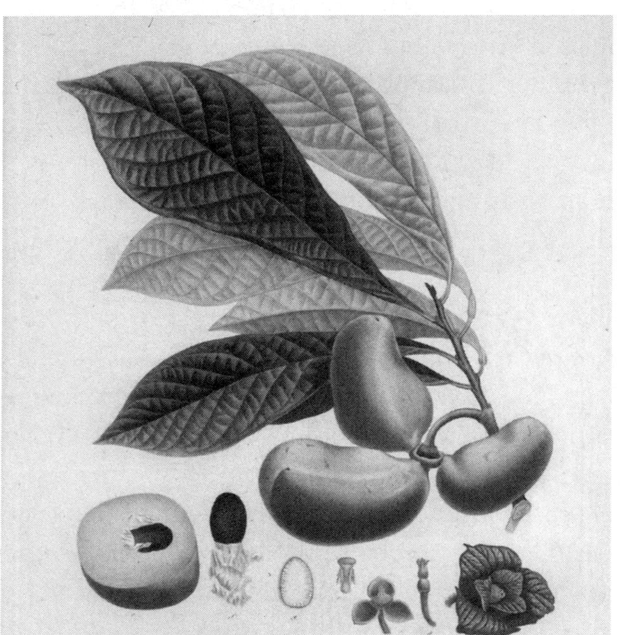

Resources

Recommended Reading

Barney, Danny, *Storey's Guide to Growing Organic Orchard Fruits*, Storey, 2013

Judd, Michael, *For the Love of Pawpaws*, Ecologia, 2019

Kentucky State University articles: uky.edu/ccd/sites; uky.edu.ccd/files /pawpaw.pdf; kysu.edu/academics/cafsss/pawpaw/pawpaw-planting-guide; kysu.edu/academics/cafsss/pawpaw/field-planting-pawpaw

Logsdon, Gene, *Organic Orcharding: A Grove of Trees to Live In*, Echo Point Books & Media, 2016

Moore, Andrew, *Pawpaw: In Search of America's Forgotten Fruit*, White River Junction, VT: Chelsea Green, 2015

Ohio Pawpaw Growers Association, *The Edible Pawpaw: A Collection of Delicious and Nutritious Recipes*

Reich, Lee, *Growing Fruit Naturally*, Taunton Press, 2012

Reich, Lee, *Uncommon Fruits for Every Garden*, 2004

SARE, *Managing Cover Crops Profitably*, USDA 1998

Supplies

A.M. Leonard: Generally the best modern mail-order source for high-quality industrial planting tools, grafting supplies, all-steel planting spades, knives, pruning and grafting tools, etc. amleonard.com

ATTRA Sustainable Agriculture: Resource base and consultancy on organic agriculture. Excellent articles and information. attra.ncat.org

Earth Tools: Great source for BCS walk-behind tractors, parts, and service and also premium-quality European hand tools, hoes, spades, loppers, etc. Located in KY. earthtools.net

E and R Seeds: Extensive catalog of growers supplies such as irrigation supplies, organic fertilizers, sprays, and greenhouses needs. Amish operated, no website; call and request a catalog (phone number can be found online). Located in IN.

eBay: Source for organic horticulture sprays, organic pesticide products, and tools.

Farm Tek and Growers Supply: Professional growers supplies, irrigations supplies, tools, parts, greenhouse kits, produce crates and boxes, growing pots, etc.

Felco Tools: Professional-quality hand pruners and loppers, saws, etc.

Fertrell: Organic bagged fertilizer mixes.

Myccorrhizal Applications: Source for professional quality fungal root inoculants.

NaturSafe: Organic bagged fertilizers.

Orchard Valley Supply: Supplier of high quality tools and orchard equipment, including ladders and Blue X Tree Tubes. orchardvalleysupply.com

Packingsuppliesbymail.com: Good source of cheap but high-quality packing supplies.

Rodale Institute: Dedicated to helping organic farming and the organic movement. rodaleinstitute.org

Seven Springs Farm: Supplier of mail-order high quality organic fertilizer and pest control products, including vegan fertilizers. Great selection. 7springsfarm.com

Uline: Packing supplies, bags, boxes, craft paper, shipping tape.

Nurseries and Groups

Baker Creek Seeds: Ultra-Select pawpaw seeds available. rareseeds.com

Peaceful Heritage Nursery: Premium-quality pawpaw trees both grafted cultivars as well as superior Ultra-Select seedlings, scion wood, and Ultra-Select pawpaw seeds. peacefulheritage.com

Pawpaw Fanatics: Facebook page for pawpaw enthusiasts and growers.

Growingfruit.org: Group forum for fruit, including pawpaw, growers and enthusiasts. Many pages devoted to pawpaw discussions.

North American Pawpaw Growers Association (NAPGA): Annual events and quarterly publications. Based in OH, and current president is Ron Powell, PhD. ohiopawpaw.com and also on Facebook.

Notes

Introduction

1. *Uncommon Fruits for Every Garden*, Lee Reich, 2004.
2. Neal Peterson adds: "The best scientific evidence has *A. triloba* growing in the temperate forest called the Arcto-Tertiary Circumpolar forest. So, actually, *A. triloba* migrated south not north."
3. nps.gov/articles/pawpaw.htm
4. Neal Peterson adds: "Another myth is that pawpaw is an aphrodisiac and euphoric. This has not been proven, and most people do not experience these effects."

Chapter 1: Foraging for Wild Pawpaws

1. "Dwarf" is a widely used horticultural term to denote a species or cultivar that is small in stature compared to the normal or standard-sized version. Examples would be dwarf apple trees (10 to 12 feet) versus standard apple tress (30 to 40 feet).
2. petersonpawpaws.com/gallery

Chapter 2: Description of North American Pawpaw Fruit

1. journals.ashs.org/hortsci/view/journals/hortsci/50/8/article-p1202.xml
2. NAPGA, *Pawpaw Pickin's*, Autumn 2013, Volume 13, Issue 2, p. 2
3. Kirk Pomper's theory on pawpaw fruit leather toxicity.
4. pawpaw.kysu.edu/PDF/AcetoUpdate3.pdf
5. pubs.acs.org/doi/abs/10.1021/np100228d
6. ncbi.nlm.nih.gov/pubmed/16078200
7. ncbi.nlm.nih.gov/pubmed/17017523
8. The largest native American tree fruit in general is the nonedible Osage orange (beaudarc, hedge-apple) (*Maclura pomifera*).
9. A peduncle is an unusual word for a specialized stem that starts out as the stem of the pawpaw flower and then, after fertilization, becomes the stem of the fruit itself.

Chapter 3: Flowering and Pollination

1. extension.missouri.edu/greene/documents/Horticulture/Presentations/PawpawMOA2_7_15.pdf
2. pawpaw.kysu.edu/PDF/CrabtreeetalThinning2010.pdf

Chapter 4: Site Design and Planting

1. KSU plants their pawpaws 8 feet apart within the rows with excellent results. Sheri Crabtree says 6 feet is a little too close together, but trees will still fruit and do all right.
2. As we noted in the moderately severe drought of summer 2019, young trees planted along the edge of large wild native trees will suffer in drought because the mature trees have enormous root systems and will use up the soil moisture rapidly, in the process dehydrating and even sometimes killing off young fruit trees, so be cautious and prepared to irrigate.
3. Anthony Bratsch, Robert Bellm, and Don Kniepkamp, "Early Growth Characteristics of Seven Grafted Varieties and Non-Grafted Seedling Pawpaw," *HortTechnology*, 13(3), July 2003.

Chapter 5: Choosing Your Trees

1. Neal Peterson.

Chapter 6: Maintaining the Orchard

1. Glyphosate deemed a "probable carcinogen" by the World Health Organization (WHO); who.int/foodsafety/faq/en/
2. kysu.edu/wp-content/uploads/2017/07/OrganicPawpawPBI-004.pdf
3. pawpaw.kysu.edu/PDF/OrganicPawpawPBI-004.pdf
4. aesl.ces.uga.edu/soil/fertcalc
5. pawpaw.kysu.edu/PDF/OrganicPawpawPBI-004.pdf
6. Moore, Andrew. *Pawpaw: In Search of America's Forgotten Fruit.* White River Junction, VT: Chelsea Green, 2015.
7. files.danfoss.com/TechnicalInfo/Dila/04/IC.PS.600.A7.02_RJA.pdf
8. pawpaw.kysu.edu/slides/PP_Slide25.htm
9. extension.missouri.edu/greene/documents/Horticulture/Presentations/PawpawMOA2_7_15.pdf

Chapter 8: Tree Propagation

1. pawpaw.kysu.edu/PDF/Pomper%20et%20al%20p145-149.pdf
2. petersonpawpaws.com/neals-story-cont
3. 'Rappahanock' pawpaw is only successful in some locations; in others it is prone to physiological problems.

Chapter 9: Pests, Diseases, Disorders, and Their Management

1. kysu.edu/wp-content/uploads/2017/07/Pawpaw-Pests-and-Diseases-2016-Final.pdf
2. NAPGA, *Pawpaw Pickin's*, Fall 2015.
3. John D. Sedlacek, Jeremiah D. Lowe, Kirk W. Pomper, Karen L. Friley, and Sheri B. Crabtree, "The Pawpaw Peduncle Borer, *Talponia plummeriana* Busck (Lepidoptera: Tortricidae): A Pest of Pawpaw Fruit," *Journal of the Kentucky Academy of Science*, 73(2), 110–112, September 2012.
4. See KSU's document, *Organic Production of Pawpaw.*
5. NAPGA, *Pawpaw Pickin's*, Volume 17, Issue 2, Spring 2017; cabi.org/isc/datasheet/57235
6. You can find instructions online for making traps, but it's doubtful these would make much difference if your trees are severely stressed; walterreeves.com/insects-and-animals/asian-ambrosia-beetle-trap/
7. entomologytoday.org/2015/03/18/study-confirms-effectiveness-of-cheap-simple-traps-for-citizen-science-project/
8. Jessica Bessin, Patty Lucas, and Ric Bessin, "Spotted Wing Drosophila and Backyard Small Fruit Production," University of Kentucky College of Agriculture; entomology.ca.uky.edu/ef231
9. Occurrence of Phyllosticta Fungal Fruit Spot and Fruit Cracking in Pawpaw (*Asimina triloba*), Poster Board #308; researchgate.net/publication/267342574_Occurrence_of_Phyllosticta_Fungal_Fruit_Spot_and_Fruit_Cracking_in_Pawpaw_Asimina_triloba_Poster_Board_308
10. NAPGA, *Pawpaw Pickin's*, Fall 2012, Volume 12, Issue 2
11. Joseph D. Postman, Kim E. Hummer, and Kirk W. Pomper, "Vascular Decline in the Oregon Pawpaw Regional Variety Trial," *HortTechnology*, 13(3), 418–420, July–September 2003.
12. NAPGA, *Pawpaw Pickin's*, Volume 17, Issue 1, 4, Spring 2017.
13. NAPGA, *Pawpaw Pickin's*, Volume 15, Issue 1, 4–5, Spring 2015.
14. ucanr.edu/sites/mgslo/newsletters/Oak_Root_Fungus28069.htm
15. Andrew Moore, *Pawpaw: In Search of America's Forgotten Fruit.*
16. NAPGA, *Pawpaw Pickin's*, Volume 16, Issue 2, 2, Fall 2018.
17. extension.missouri.edu/greene/documents/Horticulture/Presentations/PawpawMOA2_7_15.pdf

Chapter 10: Pawpaw Fruit Marketing Strategies

1. southcenters.osu.edu/sites/southc/files/site-library/site-documents/HORT/Fruits/pawpaws/PawpawProcessing.pdf
2. kysu.edu/academics/college-acs/school-of-ace/pawpaw/recipes-and-uses.php

Chapter 12: Troubleshooting, Cost Analysis, and Calendar

1. uky.edu/ccd/sites/www.uky.edu.ccd/files/pawpaw.pdf
2. kysu.edu/academics/cafsss/pawpaw/field-planting-pawpaw/
3. Cheryl Kaiser and Matt Ernst, *Pawpaw*, University of Kentucky, 2018; uky.edu/ccd/sites/www.uky.edu.ccd/files/pawpaw.pdf

Chapter 13: Conclusion

1. Bock Bioscience in Germany, through the work of Stephan Von Rundstedt and Maria Blondeau; biooekonomie.de/en/helping-pawpaw-breakthrough

Index

Cultivar index follows main index.

A

accessibility, of planting site, 49
acetogenins, 24–25
aftertaste, 22
animals, 70, 108, 118–121, 186
anthers, 32–33
ants, 116
aphids, 116–117
arils, 21–22
Armillaria root rot, 123–124
aroma, of flowers, 11
Asian ambrosia beetle (*Xylosandrus crassiusculus*), 90, 112–113, 184–185
Asimina pygmaea, 13–14
Asimina species, 13–14
Asimina triloba. *See* pawpaw
Asimina webworm (*Ompalocera munroei*), 113–114
autumn maintenance, 193

B

barbeque sauce, 144
bare-root trees, 51, 67
bark
  animal pests, 119–120
  diseases, 122
  insect pests, 117
  lesions, 125
  mechanical damage, 127–128
  sun injury, 125–126, 127
  troubleshooting disorders, 185, 186–187
bark inlay graft, 104
BCS walk-behind tractors, 69
Belgium, 159
bindweed (*Convolvus arvensis*), 76–77
black plastic mulch, 74–75
black spot (*Diplocarpon* sp.), 124, 184
black walnut (*Juglans nigra*), 72
blackberry (*Rubus* sp.), 77
Blandy Experimental Farm, 105–106, 174, 180
blue stem disease (BSD), 122, 184–185
Blue-X grow tubes, 55, 57
branches, 88–91, 186
breeding programs, 5, 24, 105–106
brewery sales, 138–141
Brinkmann-Landwehr, Hans, 160
brown marmorated stink bug (*Halyomorpha halys*), 115–116
bucket tree shelter, 58
Buckman, Benjamin, 179
buds, 31–32
bush hog damage, 127–128

C

California, 169–170
Campbell, R. Douglas, 165
Casa Nueva Restaurant, 144
caterpillars, 114
cats, 120–121
cedar, 72
Cheatwood, J., 179
Cheely, J., 179
chemical fertilizers, 80
chicken manure, 79, 80
chicken wire cage tree shelter, 56, 58
chip budding, 103
Chmiel, Chris, 70, 156, 157, 166
clay soils, 45
clingstone, 22
commercial growers, harvest season, 34
compost tea, 92
consumers, preferred fruit textures, 19
cookies, 147
cost analysis, 188–191
cover crops, 48
cream pie, 147

Creech, John V., 176
Croatia, 168
Cullman, Don, 157
cultivars
  availability of, 59, 151–152
  characteristics, 152–155
  conversion by grafting, 104–105
  development of, 105–106
  existing, 155–176
  missing/extinct, 179–182
  numbered/unreleased, 177–179
  selection, 62–63
custard, 146
custard pie, 145

D

Davis, Corwin, 156, 157, 163, 168, 169, 174, 175, 181
Davis, Lester, 179
deer, 119, 186
dehydration, 184
diseases, 10–11, 108, 121–125, 183–188
disorders, 108, 125–130, 183–188
dogs, 120–121
Dormant bloom stage, 31
Downing, Ernest J., 164, 181
drip irrigation emitters, 82–83, 87–88

E

The Edible Pawpaw (OPGA), 142
Endicott, G., 180
England, Cliff, 156, 165, 172
ethnic markets, 149

F

Fairchild, David, 180
farmer's markets, 131–133
fertigation, 81
fertilization, 78–81, 186, 187
fish emulsion, 79, 81, 92
5-gallon bucket tree shelter, 58
flame weeders, 70
flavor, 17–18, 20
flies, 34–36
flowers
  aroma, 11
  description of, 29
  frost damage, 128–129
  insect pests, 109–110
  stages of bloom, 26–27, 29–34
  troubleshooting disorders, 187–188
foliage
  damage, 130
  deformation, 123
  diseases, 121, 124
  insect pests, 110–115, 117–118
  troubleshooting disorders, 183–188
foliar spray application, 30
frass, 109, 112–113
freestone, 21
freeze, effect of, 31–33
Friedman, Carol, 159
frost, 9–10, 30–33, 128–129, 187–188
fruit
  aftertaste, 22
  animal pests, 118–119, 120–121
  diseases, 121–123
  flavor, 17–18, 20
  fruitlets, 33, 40
  fused, 129
  grades, 154–155
  growth habit, 25–26
  on immature trees, 103–104
  insect pests, 115–116
  ripening and harvesting, 40–41, 95, 153
  seed removability, 21–22
  seed-to-pulp ratio, 21
  side effects of eating, 22–24
  splitting, 130
  storage, 96
  sun injury, 126–127
  texture, 18–19, 20–21
  thinning, 38–40, 93–94
  troubleshooting disorders, 185–186, 187–188
fruit leather, 24
fruiting buds, 29
Full Bloom stage, 32–33
fungal diseases, 121–124, 184–186. See also *Phyllosticta*
fungicides, 107

G

Gable, J., 180
Gibson, Milo, 164, 172
Glaser, Dick, 171
Glaser, R., 180
goats, 70
Gordon, John, 167, 170, 171
grades, 154–155
grafted trees, 63–65, 128, 186
grafting, 103–105
grass, 69–77, 114, 183
Green bloom stage, 32
grocery stores, 134–135
ground cloth, 73–74

H

habitat, 7, 9, 14, 25
Hale, George, 159
Halvin, Tyler and Dana, 157, 159
hand-pollination, 33, 36–37

hardiness, 25, 30
harvest period, 34
harvesting, 39, 40–41, 93–96, 190–191
health sprays, 92
herbicides, 70
Hickman, Joseph W., 165
holes, for planting, 53
Hoopes, W., 180
Hope, A., 180
Horn, Al, 160
hornworms (*Manduca* sp.), 118
hybrids, 105
hydrants, 83, 87, 88

I

ice cream, 148
insects
  pests, 107–108, 109–118
  pollinators, 27, 33, 34–36
  troubleshooting pests, 183–188
interveinal chlorosis, 130
irrigation, 48–49, 82–88
Italy, 158–159, 160, 168, 172

J

Jacobs, Homer, 174
Japanese beetle (*Popillia japonica*), 112
Japanese honeysuckle (*Lonicera japonica*), 77
Johnson, Mary Foos, 164
Johnson grass (*Sorghum halepense*), 76
June drop, 93–94

K

kelp extract (*Ascophyllum nodosum*), 30, 51, 79, 81, 92
Kentucky, pawpaw habitat in, 14
Kentucky State University (KSU)
  breeding program, 5, 24, 105–106, 177–178
  on fertilizer, 78, 79
  healthy growth rates, 81
  irrigation recommendations, 85
  paint, 126
  planting density, 188
  production costs, 190
  recipes, 145–148
  thinning fruit, 39, 94
  topping trees, 89
  vole damage, 119
Ketter, F., 180
Kruszewski, Oriana, 158, 160
Kurle, R., 180

L

lassi, 144
latex paint, 120, 126, 185, 186–187
lawnmower damage, 127–128
leaves. *See* foliage
Lecanium scales, 116
Lehman, Jerry, 106, 156, 161, 163, 164, 166, 169, 170, 178–179
lifespan, 66
limbs, 88–91, 186
Little, J. A., 181
loam soils, 46
Louisiana State University, 156
Lowe, Jeremy, 162, 177

M

magnesium deficiency, 187
magnolia scales (*Neolecanium cornuparvum*), 116
mail orders, 135–137
maintenance, yearly, 191–193
Mansell, Tom, 161
market growers, flowering cycle length, 34
marketing strategies
  consumer preferences, 19
  ethnic markets, 149
  farmer's markets, 131–133
  grocery stores, 134–135
  online orders, 135–137
  pulp sales, 141–142
  recipes, 144–148
  restaurant and brewery sales, 138–141
  U-pick farms, 137–138
  value-added goods, 142–143
Mckay, John W., 180, 181
mechanical damage, 127–128, 185
mice, 119–120
milkshake, 144
Missouri State University, 34
modified bark grafting, 103
Montanari, Domenico, 160, 168
mowing, 69–71
mulch, 70, 71–76
Munich, Don, 172
mycorrhizal inoculant, 52

N

Native Americans, 6
nausea, 22–23
Neighbors, Ken, 168
Norris, Sam, 170
North American Pawpaw Growers Association, 17
North American pawpaw, use of term, 17
northern flatid planthopper (*Flatormensis proxima*), 115
nutrient deficiencies, 183, 186, 187–188

O

oak root fungus, 123–124
online orders, 135–137
opossums, 119
orange-fleshed varieties, flavor, 20
orchard
  annual maintenance, 191–193
  choosing cultivars, 61–63
  converting cultivars by grafting, 104–105
  cost analysis, 188–191
  older trees, 186
  planting, 50–55
  site design, 49–50
  site selection, 43–49
  weed and grass control, 69–78
organic fertilizers, 80
organic mulches, 72–73
Oswald, E., 181

P

Pacific Northwest, 122, 184–185
paint, 120, 126, 185, 186–187
papaya (*Carica papaya*), 8, 9, 17
parfait, 146
pawpaw, cultivated varieties (*Asimina triloba*)
  availability of, 2
  breeding programs, 5, 24, 105–106
  future of, 195–196
  growth characteristics, 25–27
  interest in, 1, 4
  myths about, 8–11
  other fruit with same name, 8
  See also cultivars
pawpaw, wild (*Asimina triloba*)
  foraging for, 3, 13, 16
  growth habit, 15
  habitat, 9, 14, 25
  history of, 5–7
  lack of fruit, 187
  range, 7–8
  transplanting, 102
Pawpaw Cookies with Black Walnuts, 147
Pawpaw Cream Pie, 147
Pawpaw Custard, 146
Pawpaw Custard Pie, 145
Pawpaw Ice Cream, 148
Pawpaw Lassi, 144
pawpaw leaf roller (*Choristoneura parallela*), 112
pawpaw peduncle borer (*Talponia plummeriana*), 109–110, 188
Pawpaw Pie or Parfait, 146
pawpaw sphinx moth (*Dolba hyloeus*), 114
peduncles, 25–26, 39, 109–110
persimmon (*Diospyros virginiana*), 10
pesticides, 107
pests
  animals, 108, 118–121
  grass growth and, 71
  insects, 107–108, 109–118
  susceptibility to, 10–11
  troubleshooting, 183–188
Peterson, Neal
  breeding programs, 105–106
  on cultivar quality, 159, 165
  cultivars, 155, 168, 169, 171, 174, 175
  drought losses, 82
  fertilization, 78, 130
  fruit leather research, 24
  grafting, 103
  hybridization work, 14
  on tetraploids, 170
Phyllosticta, 56, 57, 121, 130, 185–186
physiological disorders, 108, 125–130, 183–188
pie, 145, 146, 147
pillbugs (*Armadillidium vulgare*), 72, 117
Pink bloom stage, 32
pink discoloration of fruit flesh, 122–123
planting
  best practices, 53–55
  potted vs. bare-root, 51
  site design, 49–50
  site selection, 43–49
  supplies needed, 51–52
  tree protection, 55–58
plastic mulch, 74–75
pollen, 33
Pollen Shed bloom stage, 33
pollination, 27, 33, 36–38, 68, 187
pollinators, 7, 27, 33, 34–36
Pomper, Kirk, 176
potted trees, 51
Potter, B. S., 180
Powell, Ron
  on cultivar quality, 159, 165, 169
  cultivars, 123, 124, 157, 160, 171
pressure reducers, 83, 84
prices, of premium fruit, 95
production costs, 190–191
propagation

- breeding programs, 5, 24, 105–106
  - goals of, 97
  - grafting, 103–105
  - seedlings, 97–102
- pruning, 88–91
- pulp production
  - choosing cultivars, 62
  - effect of aril, 22
  - harvesting, 95, 96
  - processing, 141–142
  - recipes, 144–148
  - value-added goods, 142–143

R

- raccoons, 118–119
- recipes, 144–148
- Rees, W., 181
- Reich, Lee, 167, 171
- restaurant sales, 138–141
- rhizopus rot (*Rhizopus stolonifer*), 121–122
- Riley, J. M., 169
- ripening, 9–10, 22–23, 40–41, 153
- Roach, J. C., 181
- rock phosphate, 52
- rocky soil, 46–47
- rollie-pollies (*Armadillidium vulgare*), 72, 117
- Romania, 172
- root system
  - diseases, 123–124
  - disturbances, 50, 183
  - seed growth, 101–102
  - smaller trees and, 59–60
- rootstock, 98, 103–104
- Russia, 176

S

- saddleback caterpillar (*Acharia stimulea*), 114
- sandy soils, 46
- saplings, UV protection, 55–58
  - *See also* grafted trees; seedlings
- sapsuckers (*Sphyrapicus varius*), 120, 185
- scales, 116
- Schmidt, Al, 144
- Schubert, K., 181
- Scott, C. S., 181
- seed characteristics in fruit, 21, 22
- seed collection and growing, 98–102
- seedlings
  - growing, 97–102
  - lifespan, 64
  - planting orchard with, 62, 66–67
  - sun injury, 55–56, 127
  - value of, 68
- self-pollination, 36, 37–38
- selling fruit. *See* marketing strategies
- sepals, 32
- Shawie, Amanda, 145
- shipping fruit, 135–137
- shrubs, growth characteristics, 26
- Siamese fruit, 129
- side effects of eating, 22–24
- site design, 49–50
- site selection, 43–49
- Slate, George, 167, 176
- slugs and snails, 114–115
- soil, 44–47, 184
- sow bugs/pillbugs, 72, 117
- spacing, 37, 49–50
- spider mites, 117–118, 184
- spotted wing drosophila (SWD) (*Drosophila suzukii*), 115
- spraying, 91–92, 107
- spring maintenance, 192
- stigmas, 32–33
- stinging rose caterpillar (*Parasa indetermina*), 114
- storage, 40–41
- Stout, W. C., 180
- stratification, 99–100
- string trimmer damage, 127–128, 185
- suckers, 77–78, 195
- summer maintenance, 192
- sun, requirements, 43–44
- sun damage
  - fruit, 126–127
  - saplings, 55–56, 127
  - troubleshooting, 184–185, 186–187
  - winter bark injury, 125–126, 127
- Sweet Hot 'n Tangy Barbeque Sauce, 144
- synthetic mulches, 73–76

T

- Talbot, John, 181
- T-budding, 103
- temperature, effect on bloom stages, 31–33
- tetraploids, 170
- texture, 18–19, 20–21
- thinning fruit, 38–40, 93

thrips, 118, 184
timers, 83–85
tissue culture, 195–196
topping trees, 89
training branches, 88–91
transplant shock, 50, 183
transplanting, 59–60, 102, 183
tree shelter systems, 55–58
trees
  grafted, 63–65
  growth characteristics, 26
  planting, 50, 53–55
  propagation, 97–106
  purchasing, 188–189
  seedlings, 65–68
  size of transplants, 59–61
  spacing, 37
  troubleshooting disorders, 183–186
triploid pawpaw, 106
trunk
  animal pests, 119–120
  diseases, 122
  insect pests, 112–113, 117
  lesions, 125
  mechanical damage, 127–128
  sun injury, 125–126, 127
  troubleshooting disorders, 185, 186–187
Tubex shelters, 55, 57

U

understory plants, 75–76
U-pick farms, 137–138
USDA, 122
UV light. *See* sun damage

V

value-added goods, 142–143
Van Der Bogart, Francis Van Der, 181
varieties, flavor differences, 20
Vegan Pawpaw Zucchini Bread, 145
Velvet Bud bloom stage, 31
Velvet Stem bloom stage, 31–32
viral diseases, 187
voles, 119–120

W

Walker, Woody, 157, 160, 161, 164, 175
Ward, W. B., 165
weed control, 71–78, 114
Wells, David, 176
wet soils, 47
whip and tongue grafting, 103
white-fleshed varieties, flavor, 20
whiteflies, 118, 184
whitewashing. *See* paint
wild mulberry (*Morus alba*), 77
wild papaws. *See* pawpaw, wild
wild rose (*Rosa multiflora*), 77
wind, 47–48
winter bark injury, 125–126, 127, 186–187
winter maintenance, 191–192
wood chips, 73

Y

yellow-bellied sapsucker (*Sphyrapicus varius*), 120, 185
yellow-fleshed varieties, flavor, 20
yields, 68, 94

Z

zebra swallowtail butterfly (*Eurytides marcellus*), 110–112
Zimmerman, G. A., 171, 176, 180, 181
zoysia grasses, 77
zucchini-pawpaw bread, 145

Cultivar Index

'1 XL', 160–161
'9-58', 122
'166-13' ('Maria's Joy'), 63, 106, 164
'166-66', 178
'250-30', 161
'250-39' ('Jerry's Big Girl'), 106, 161
'275-17', 166
'275-25', 178
'275-48' ('Lehman's Delight'), 106, 163
'275-50', 178
'275-56', 178
'275-56N', 179
'275-56S', 179
'275-60', 179
'275-69', 179
'400-27', 179
'400-50', 179
'400-275', 179
'Al Horn Whiteflesh', 160
'Ark-21', 156
'Arkansas Beauty', 179
'Atria', 156

'Baker's Acres #1', 156
'Baker's Acres #2', 156
'Belle', 156
'Benny's Favorite' ('VE-21'), 106, 156
'Bertria', 156
'Betty Wirt', 179
'Broad', 156
'Buckman', 179

'Cale's Creek', 156, 179
'Cantaloupe', 156–157
'Car Wreck', 157
'Catochin', 156
'Cawood', 156
'Cheatwood', 179
'Cheely', 179
'Cherokee Ridge', 157
'Collins', 157
'Cox's', 157
'Cox's Favorite', 180
'Creek Ridge', 157
'Cullman Late', 157

'Danae's Creekside', 157
'Danika', 157
'Davis', 157
'Dr. Chill', 158
'Dr. Potter', 180
'Duck', 180

'Early Best', 180
'Early Cluster', 180
'Early Gold', 180
'Early Surprise', 158
'Endicott', 180
'Estil', 158

'Fairchild', 180
'Fairchild #2', 180
'Ford Amend', 158, 185
'Fort Hunter', 158
'Franklin Valley Gold', 158

'G-2', 180
'G6-120' ('Jeremy's Gold'), 161, 177
'G9-108', 177
'G9-109', 177
'G9-111', 177
'Gable', 180
'Gainesville #1' and '#2', 158
'Gene', 158
'Georgia', 158–159
'Ghent 2 Blom', 159
'Glaser', 158
'Golden Moon', 40, 159
'Greenriver Belle', 159

'H-3-120' ('State Fair'), 172, 177
'Hale #5', 159
'Halvin', 40, 62, 159–160
'Halvin's Danae's Creekside', 157
'Hann', 180
'Hardy Wonder', 160
'Hayes', 160
'Haz-1', 177
'Hengst', 180
'Hoberg', 160
'Holt', 180
'Holtwood', 180
'Honey Dew', 160
'Hoot Owl', 160
'Hope's August', 180
'Hope's September', 180
'Hopi Ridge', 160
'Horn's White', 160

'Ithaca', 160
'IXL', 160–161

'Jack Irvine #1', 161
'Jenny's Gold', 161
'Jeremy's Gold' ('G6-120'), 161, 177
'Jerry's Big Girl', 106, 161
'Jerry's Delight', 161
'Jonathan', 161
'Jumbo', 180

'K 8-2'. *See* 'KSU Atwood'
'Kentucky Champion', 161
'Kercheval', 180
'Ketter', 180
'Kings Gold #2', 161
'Kirsten', 161
'KSU 2-11', 177
'KSU 27', 177
'KSU Atwood' ('KSU 8-2'), 63, 64, 106, 162, 178, 185
'KSU Benson' ('KSU 7-5'), 63, 106, 162
'KSU Chappell' ('KSU 4-1'), 106, 162–163
'KSU G4-25' ('Pina Colada'), 178
'KSU G6-120', 161
'KSU HI-1-4', 177
'KSU HI-7-1', 177
'Kurle', 180

'Lady D', 163
'Lawvere', 180
'Lehman's Chiffon', 163
'Lehman's Delight', 106, 163
'Little Rosie', 180
'Long John', 181
'Louisiana Native', 163
'Lynn's Favorite', 163

'M. Gordon', 164
'M-1', 181

'Mammoth', 163
'Mango', 163–164
'Maria's Joy', 63, 106, 164
'Mark Twain', 164
'Martin', 181
'Mary Foos Johnson', 164
'Mason-WLW', 181
'Middletown', 164–165
'Mitchell', 165
'Mohican', 165
'Montanari 1216', 168
'Mudge', 181

'NC-1', 40, 62, 63, 165
'Neal', 165
'NN10-35', 173
'Nyomi's Delicious', 40, 165–166

'October Moon', 166
'Osbourne', 181
'Oswald', 181
'Overleese', 63, 106, 165

'Pennsylvania Golden' series, 40, 62, 166–167
'Pepper 1', 166
'Pepper 2', 166
'Pickle', 166
'Pina Colada' ('KSU G4-25'), 178
'Potomac', 63, 106, 167–168
'Prairie King #1', 168
'Prairie King #2', 168
'Prima 1216', 168
'Probst Early', 181
'Prolific', 168–169

'Quaker Delight', 169

'Rana', 169
'Rappahannock', 106, 169
'Rebecca's Gold', 169–170
'Rees', 181
'Regulus', 170
'Rigel', 170
'Roach', 181
'Ruby Keenan', 170

'SAA Overleese', 170
'SAA Zimmerman', 171
'SAB Overleese', 170
'SAB Zimmerman', 171
'SAC Overleese', 170–171
'SAC Zimmerman', 171
'Sam Norris' cultivars, 170
'Sarah G', 171
'Schriber', 181
'Scott', 181
'Seneca Ridge', 171
'Shannondale', 181
'Shawnee Trail', 171
'Shenandoah', 63, 171–172
'Sibley', 172
'Silver Creek', 181
'Simina', 172
'State Fair' ('H-3-120'), 172, 177
'Sue', 172
'Summer Delight', 153, 172
'Sunflower', 22, 63, 106, 172–173
'Sun-glo', 172
'Super Mario', 173
'Susquehanna', 22, 62, 63, 106, 173
'Sweet Alice', 174
'Sweet Potato', 174
'Sweet Virginia', 174

'Talbot', 181
'Tallahatchie', 106, 174
'Taylor', 174, 181
'Taytwo', 174–175
'Tollgate', 175
'Tropical Treat', 175

'Uncle Tom', 181
'UVM #1', 175

'Van Der Bogart', 181
'VE-5', 178
'VE-9', 178
'VE-21' ('Benny's Favorite'), 106, 156
'Vena', 181
'Vickey Russell', 175
'Victoria', 175

'Wabash', 106, 175–176
'Walter's', 176
'Wells' ('Well's Delight'), 176
'Whip Longsderf', 176
'White Owl', 176
'Wilson', 122, 154, 176, 185

'Yuri's Russian', 176

'Zimmerman', 176

About the Author

BLAKE COTHRON is an organic farmer, professional nurseryman, musician, poet, writer, educator, and horticulturist with over 20 years' experience in organic agriculture, botany, horticulture, and growing food. Blake owns and operates Peaceful Heritage Nursery, a four-acre USDA Certified Organic research farm, orchard, and edible plant nursery, with a special focus on pawpaws. He shares his two decades of experience in organic agriculture and horticulture through magazine articles, including in Permaculture Design, online publications, public speaking engagements, a Youtube channel, and blogging. He splits his time between farming, research, gardening, travel, yoga, meditation, and being a husband and father. He lives with his wife and son in beautiful Kentucky.

ABOUT NEW SOCIETY PUBLISHERS

New Society Publishers is an activist, solutions-oriented publisher focused on publishing books for a world of change. Our books offer tips, tools, and insights from leading experts in sustainable building, homesteading, climate change, environment, conscientious commerce, renewable energy, and more—positive solutions for troubled times.

We're proud to hold to the highest environmental and social standards of any publisher in North America. When you buy New Society books, you are part of the solution!

DON'T EAT THIS BOOK (but you could)

- We print all our books in North America, never overseas
- All our books are printed on **100% post-consumer recycled paper**, processed chlorine-free, with low-VOC vegetable-based inks (since 2002)
- Our corporate structure is an innovative employee shareholder agreement, so we're one-third employee-owned (since 2015)
- We're carbon-neutral (since 2006)
- We're certified as a B Corporation (since 2016)

At New Society Publishers, we care deeply about *what* we publish—but also about *how* we do business.

To download our full catalog, please visit newsociety.com/pages/nsp-catalogue

ENVIRONMENTAL BENEFITS STATEMENT

New Society Publishers saved the following resources by printing the pages of this book on chlorine free paper made with 100% post-consumer waste.

TREES	WATER	ENERGY	SOLID WASTE	GREENHOUSE GASES
47	3,700	20	160	20,200
FULLY GROWN	GALLONS	MILLION BTUs	POUNDS	POUNDS

Environmental impact estimates were made using the Environmental Paper Network Paper Calculator 4.0. For more information visit www.papercalculator.org.